Better Construction Briefing

"In *Better Construction Briefing*, we finally have a much needed insight into the dynamics of briefing. Grounded on thorough research this book explores the briefing process from the perspectives of the key groups involved in it. This exposition, unlike many other books on the subject, focuses on the needs of all those involved. Indeed it provides a very human perspective of briefing, and recognises that briefing is much more than simply one of following a prescribed sequence of mechanistic steps. Barrett and Stanley's book is well structured and easy to follow, with very clear action orientated summaries enabling the reader to take some of the ideas and put them into practice. In particular the case studies at the end of the book provide very good examples of real-life projects where these ideas have been implemented."

Matthew Bacon
Head of Implementation Strategy, BAA

"*Better Construction Briefing* is a welcome contribution to achieving best practice. It provides an excellent insight into the importance of customers throughout the procurement process and how best to deliver a product, which fully meets their needs. An excellent example of collaboration between academia and practice."

Trevor Mole
Property Tectonics
President of the Building Surveyors division, the Royal Institution of Chartered Surveyors

"This excellent book is an important addition to the field of briefing/programming. No other book presents such a thorough research-based assessment of the barriers to effective briefing and practical ways to overcome them. Poignant illustrations and case studies help the reader visualize the barriers to briefing, what to do to improve the briefing process, and how to go about implementing the improvements in your own practice."

John Zeisel
President, Hearthstone Alzheimer Care

Better Construction Briefing

Peter Barrett MSc, PhD, FRICS
and
Catherine Stanley BA, DipArch
The University of Salford

Blackwell Science

Editorial Offices:
Osney Mead, Oxford OX2 0EL
25 John Street, London WC1N 2BL
23 Ainslie Place, Edinburgh EH3 6AJ
350 Main Street, Malden
MA 02148 5018, USA
54 University Street, Carlton
Victoria 3053, Australia
10, rue Casimir Delavigne
7006 Paris, France

Other Editorial Offices:

Blackwell Wissenschafts-Verlag GmbH
Kurfürstendamm 57
10707 Berlin, Germany

Blackwell Science KK
MG Kodenmacho Building
7–10 Kodenmacho Nihombashi
Chuo-ku, Tokyo 104, Japan

First published 1999

Set in 11/13pt Bembo
by DP Photosetting, Aylesbury, Bucks
Printed and bound in Great Britain by
MPG Books Ltd, Bodmin, Cornwall

DISTRIBUTORS

Marston Book Services Ltd
PO Box 269
Abingdon
Oxon OX14 4YN
(*Orders:* Tel: 01235 465500
Fax: 01235 465555)

USA
Blackwell Science, Inc.
Commerce Place
350 Main Street
Malden, MA 02148 5018
(*Orders:* Tel: 800 759 6102
781 388 8250
Fax: 781 388 8255)

Canada
Login Brothers Book Company
324 Saulteaux Crescent
Winnipeg, Manitoba R3J 3T2
(*Orders:* Tel: 204 837 2987
Fax: 204 837 3116)

Australia
Blackwell Science Pty Ltd
54 University Street
Carlton, Victoria 3053
(*Orders:* Tel: 03 9347 0300
Fax: 03 9347 5001)

A catalogue record for this title is available from the British Library

ISBN 0-632-05102-7

Library of Congress
Cataloging-in-Publication Data
Barrett, Peter, professor.
Better construction briefing/Peter Barrett and Catherine Stanley.
p. cm.
Includes bibliographical references.
ISBN 0-632-05102-7 (pbk.)
1. Construction industry–Communication systems. 2. Building–Superintendence. 3. Communiction of technical information. 4. Sales presentations. I. Stanley, Catherine. II. Title.
TH215.B37 1999
690′.068′8–dc21 98-51418
CIP

For further information on
Blackwell Science, visit out website
www.blackwell-science.com

Contents

Preface

This book addresses a key topic for construction clients and their professional advisors. Construction briefing is defined as the process running throughout the construction project, by which means the client's requirements are progressively captured and translated into effect. It is argued in Chapter 1, on the basis of very little progress in the last thirty years, that there is no simple 'cookbook' solution to good briefing, but that there are clear areas where improvement effort can be productively invested. Hence the emphasis in the title on *improvement* and the identification early in the book of five key improvement areas. These are: empowering the client, managing the project dynamics, appropriate user involvement, appropriate team building and the use of appropriate visualisation techniques. The 'five box' improvement model is a key contribution to the theory and practice of construction briefing.

The book explicitly deals with the problems of implementing new ideas in the pressured world of practice. A clear improvement process is described in Chapter 7 which, however, demands that firms adapt it to their own needs and priorities. The essence of the approach is to start on a limited, achievable basis on something that is important to the firm and by sustained, incremental effort this leads inevitably to wide-ranging radical change. In order to support this improvement process, the book is presented in an innovative format to make it as simple and risk free as possible for clients and those in the industry to take action. In practical terms this means that in Chapters 2–6 the five improvement areas are described in detail with extensive quotations from clients and practitioners in the margins to bring the issues to life. Conversely in Chapters 8–11 real life case study stories of implementing the ideas are told, and in the margin a parallel commentary links the action back to the five improvement areas.

It is shocking that briefing is still such an intractable problem in so many projects. As a key part of the relationship between clients and the industry, under-performing on briefing has many adverse implications. Some are short-term, such as inefficiencies owing to 'late' changes and annoyance on the client's part because of the industry's reluctance to understand their needs. Others are longer term, such as inefficiencies and ineffectiveness in the client's operations over the years that follow, owing to buildings that do not meet their full requirements. There is also the long-term effect on the business success of the companies in the industry. Satisfying clients is a key requirement for longer term success. It is shown in Chapter 9 how efforts to improve briefing practice in the five areas also actively market the construction firm in the most positive way possible. Good briefing is good business. To ignore this is business suicide, as other clients and construction companies will be actively improving their briefing practices and raising expectations for everyone.

There is no longer an excuse for complacency. This book sets out a focus for change around the 'five box' model, it provides a practical improvement process for firms to adapt to their specific circumstances and, lastly, it provides both clear ideas and numerous practical examples so that those in practice can choose what they want to emphasise – and take action now!

1
Introduction and focus

1
Introduction and focus

Aim of the book

We imagine the reader of this book, be they a construction professional or a construction client, as someone who knows the briefing process could be done better, but wonders where to start. They have possibly looked at good practice suggestions, but not found them sympathetic with their current practices.

stimulate sustained, better briefing

This book aims to give such people, with a desire to improve their briefing practice, three things to stimulate sustained, better briefing. First, a choice of key improvement areas known to impact positively on a wide range of problem areas. Second, a large number of selected examples from other people's experience. Third, an explicit, realistic, step-by-step change process that those wanting to make the journey can follow.

Problem

Ministry of Public Buildings and Works (1964) *The Placing and Management of Contracts for Building and Civil Engineering Work*, (The Banwell Report), HMSO, London.

Latham, M. (1994) *Constructing the Team*, (The Latham Report), HMSO, London.

In 1964 the Banwell Report stated that insufficient resources were devoted to defining project requirements and that this led to many problems in the construction process thereafter. Thirty years later, the 1994 Latham Report concluded that, amongst other things, more effort was required to understand clients' needs. Clearly the briefing process is seen as both critical to successful construction yet problematic in its effectiveness. Furthermore, this is a particularly intractable problem that does not seem to go away.

Many people in construction feel the brief is not a problem. A conversation is had with the client, their requirements are written down and the brief is thus created. It may be a detailed volume or perhaps just a few pages, or it may possibly be a collection of letters in the correspondence file. In any event the 'brief' then sits on the shelf and is not a problem to anyone – but neither is it relevant to anyone. It has become a side issue that may only be referred to if there is a problem. From the client's perspective this sort of brief can feel like a cage. It is constructed quickly and efficiently at the start of the project. If any attempt is made to change the parameters of the project, as the client's confidence, knowledge and feel for the issues increase, the client is politely reminded of their original statements and safely returned to their cage! Not surprisingly the client involved in a project of this sort is often disappointed with the building finally produced. The industry may have taken on all of the problems for them, but a full understanding of the client's real underlying needs has never been allowed to surface. At best the outcome can be efficient, but it stands very little chance of being effective in meeting the client's needs.

. . . the client is politely reminded of their original statements and safely returned to their cage!

The implication is that briefing must be seen as a process not an event. In addition, it is a process that not only starts early, but continues to inform all the technical work throughout the project. This is highlighted in an approach to briefing drawn from The Netherlands and shown in Figure 1.1. Here the brief is explicitly managed to evolve through various stages in parallel with the technical information. Continued interaction with the client is essential to this process, the underlying principle of which is to decide as little as possible at each stage! This means identifying the critical decisions and addressing these, but leaving flexibility on other issues for later consideration as more information becomes available. This thinking can then be extended right through the construction stage so that, as myriad specific choices are taken, the framework of the brief is used to ensure that they make global as well as local sense.

briefing must be seen as a process not an event

How briefing is conceived of is therefore very important. Seen as 'the brief', limited impact is possible. It can capture the client's initial idea, although this will usually be overshadowed by the industry's strong ownership of its design contribution. Even if the idea survives, it is like innovation without implementation. However, seen as the process running throughout the construction project by which means the client's requirements are progressively captured

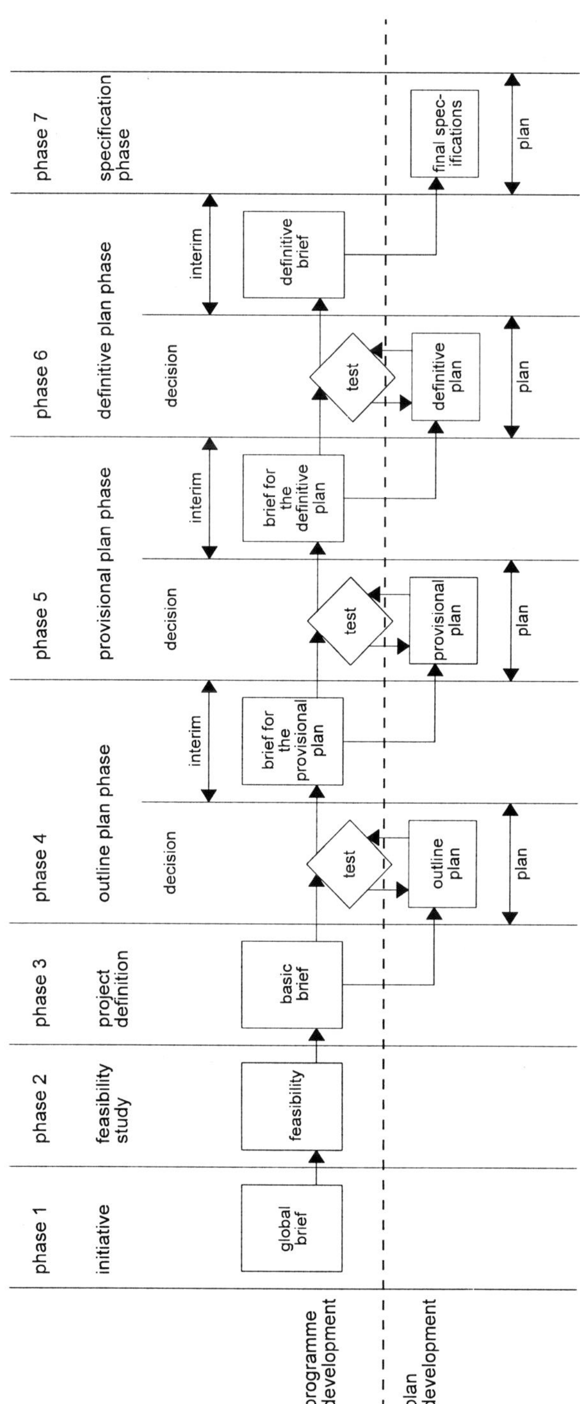

Figure 1.1
Briefing as a process

and translated into effect, briefing is central to a successful project. What must be considered is why good practice advice over the last thirty years has failed to create any significant improvement in briefing practice.

. . . why good practice advice over the last thirty years has failed to create any significant improvement in briefing practice.

What is the role of 'good practice' advice?

Since the 1960s a considerable body of work on the problems of briefing for construction projects has emerged. This has ranged from academic research reports such as Newman *et al.* (1981) and Mackinder and Marvin (1982) through to practical guides for practitioners such as Pena *et al.* (1977), O'Reilly (1987) and Salisbury (1990).

A selected, annotated bibliography of a few particularly useful sources is provided in Appendix A. So, given this wealth of information and advice, why have problems persisted?

Newman, R., Jenks, M., Bacon, V. and Dawson, S. (1981) *Brief Formulation and the Design of Buildings*, Oxford Polytechnic, Oxford.

Mackinder, M. and Marvin, H. (1982) *Design Decision Making in Architectural Practice*, Institute of Advanced Architectural Studies: Research Paper 19, University of York, York.

Pena, W.M., Caudill, W. and Focke, J. (1977) *Problem Seeking: an architectural programming primer*, Cahners, Boston.

O'Reilly, J. (1987) *Better Briefing Means Better Buildings*, Building Research Establishment, Garston, Watford.

Salisbury, F. (1990) *Architect's Handbook for Client Briefing*, Butterworth Architecture, London.

Use of good practice advice

A study of five projects threw some light on this mystery. Many problems were found. However, several of the organisations consulted had developed their own 'good practice' approaches to handling certain problems, sometimes drawing from generally accepted good practice recommendations that already exist in relation to briefing. Despite these attempts to overcome known problems, there were numerous occasions when the implementation of such ideas failed to produce a satisfactory outcome for some or all of the parties involved. Table 1.1 highlights representative examples of problems and associated attempts at solutions, termed 'potential good practice', together with the main reason for failure in each case. When analysed in the context of published good briefing practice, it was found that the reasons for the failure could be divided into three categories. These are indicated by the shading of the boxes in Table 1.1.

The data and analysis underpinning this section and much of the rest of the book are drawn from a major thirty-month research project that brought together researchers from Salford University's five-star rated Research Centre for the Built and Human Environment and a range of industrial collaborators. Further details of this work and the methods used are given in Appendix C

Table 1.1
Briefing problems and good practice

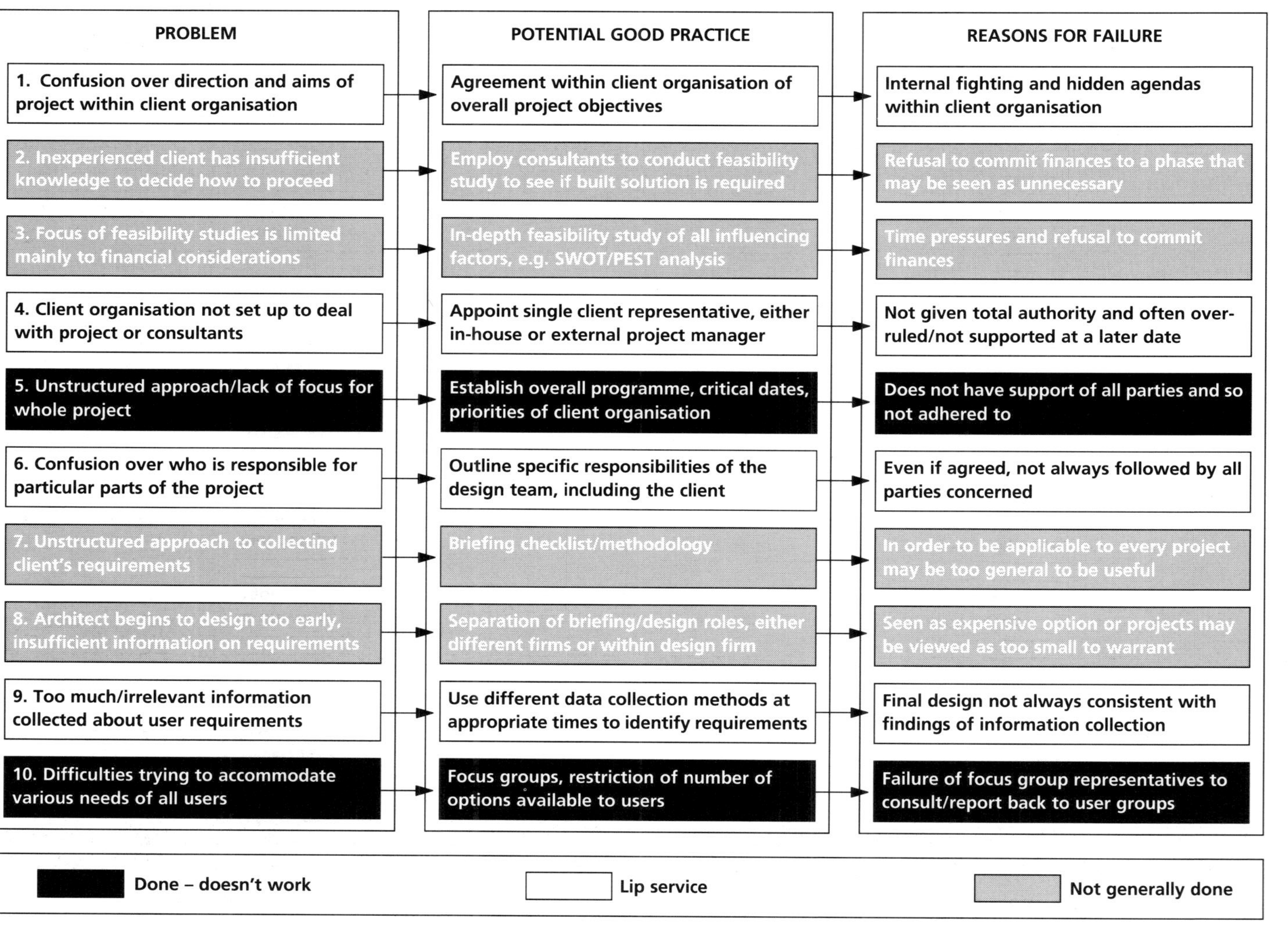

PROBLEM	POTENTIAL GOOD PRACTICE	REASONS FOR FAILURE
1. Confusion over direction and aims of project within client organisation	Agreement within client organisation of overall project objectives	Internal fighting and hidden agendas within client organisation
2. Inexperienced client has insufficient knowledge to decide how to proceed	Employ consultants to conduct feasibility study to see if built solution is required	Refusal to commit finances to a phase that may be seen as unnecessary
3. Focus of feasibility studies is limited mainly to financial considerations	In-depth feasibility study of all influencing factors, e.g. SWOT/PEST analysis	Time pressures and refusal to commit finances
4. Client organisation not set up to deal with project or consultants	Appoint single client representative, either in-house or external project manager	Not given total authority and often over-ruled/not supported at a later date
5. Unstructured approach/lack of focus for whole project	Establish overall programme, critical dates, priorities of client organisation	Does not have support of all parties and so not adhered to
6. Confusion over who is responsible for particular parts of the project	Outline specific responsibilities of the design team, including the client	Even if agreed, not always followed by all parties concerned
7. Unstructured approach to collecting client's requirements	Briefing checklist/methodology	In order to be applicable to every project may be too general to be useful
8. Architect begins to design too early, insufficient information on requirements	Separation of briefing/design roles, either different firms or within design firm	Seen as expensive option or projects may be viewed as too small to warrant
9. Too much/irrelevant information collected about user requirements	Use different data collection methods at appropriate times to identify requirements	Final design not always consistent with findings of information collection
10. Difficulties trying to accommodate various needs of all users	Focus groups, restriction of number of options available to users	Failure of focus group representatives to consult/report back to user groups

Done – doesn't work Lip service Not generally done

PROBLEM	POTENTIAL GOOD PRACTICE	REASONS FOR FAILURE
11. Lack of management interest, client representative expected to get on with it	Keep management informed of progress to maintain commitment to project	Difficult to maintain interest because core business seen as a priority, not project
12. Underemployment and inappropriate timing of consultants, i.e. too late	Greater involvement of all consultants during briefing and design steps	Architect/project manager may resent interference during early briefing stages
13. Briefing information still being given during late design and construction stages	Fixing of brief	Impossible expectation because further information nearly always surfaces later
14. Inability of client to read drawings and understand construction jargon	Variety of media to ensure client fully understands the scheme	Not widely used – e.g. simulation seen as too expensive
15. Client does not verify final design, so tender information may be incomplete	Client should always check final design prior to tender to ensure it relates to brief	Client under time pressure, overreliance on professionals to interpret requirements
16. Contractor has no real understanding of client objectives	Relevant briefing information should be included in the tender document	Unclear if such information would be useful to contractors
17. Failure to utilise contractors' expertise	Use of contractors to identify potential cost/time/design improvements	Lack of financial incentive to carry out extra work
18. Client changes introduced during construction	Risk minimised if majority of above suggestions followed, but may still occur	Impossible to prevent client changes
19. Disputes over decisions made during project which may lead to litigation	Ensure complete documented record of all stages of the project	Failure of different parties to maintain proper/accurate records of project
20. On project completion tend to forget problems, rather than learn from them	Formal review procedure	Attempted in half-hearted manner, without participation of all groups

Key: Done – doesn't work (black) | Lip service (white) | Not generally done (grey)

- Done – doesn't work: row 11
- Lip service: rows 14, 15, 18, 19
- Not generally done: rows 12, 13, 16, 17, 20

In the first category good practice recommendations have been *followed* which should, theoretically, have led to success, but, in practice, may not have led to a successful outcome because other external factors have not been considered. An example of this is the use of focus groups to improve building user group participation in briefing. It has been found that this may falsely raise expectations leading to disappointment later when people realise that their views have not been fully incorporated.

In the second category good practice recommendations are paid *lip service* but are only partially implemented. An example of this is the appointment of a single client representative. Such representatives are not always given total authority to make decisions and are often overruled at a later date.

In the third category are good practice recommendations that are *not generally implemented.* An example of this is the recommendation that clients should always check the design against the brief to ensure that their requirements are met. This is not generally done as clients are often under time pressure and therefore tend to rely too heavily upon professional advisors to interpret their needs. However, during early analysis of the case studies it was found that cases of 'bad practice' running counter to received wisdom have not always led to poor project performance as perhaps would be expected. An example of this is where an architect failed to carry out a proper survey of a building during a refurbishment scheme. This meant that the scheme layout that had been approved by the client was unachievable. As a result the contractor had to re-design many aspects of the work as the project progressed in consultation with the client/users. Despite this the site team rose to the challenge and the work was completed on time and the client was fully satisfied.

'bad practice' running counter to received wisdom has not always led to poor project performance

A summary of the possible outcomes of implementing or failing to implement recognised good practice is contained in Figure 1.2.

It can clearly be seen that the situation is much more complex than, simply, for known 'good practice' to be implemented. Good practice that does not work, and 'bad practice' that does, must be accounted for and the damaging effects of paying lip service understood and avoided. This calls for digging further beneath the surface.

Action		Result / Impact +ve	-ve
Good practice	Fully implemented	✓	✓
	Lip service		✓
Bad practice		✓	✓

Figure 1.2
The impact of good and bad practice

The human dimension

Further examination of the twenty examples of reasons for briefing failure (Table 1.1) shows that human nature is often at the root of these failures. Although there was often a desire to follow good practice ideas, in very few examples were the methods fully implemented as people tended to overrule part of the process. Thus the human dimension is critical and the impact of people's thinking upon the briefing process needs to be considered in more detail.

in very few examples were the methods fully implemented

When considered from the point of view of the construction professional, briefing would appear to be a very personal undertaking. It is about effective communication – getting inside the head of the client – and so there is really no right or wrong way of achieving this. It is clear from the case study data that over the years people find a particular method for eliciting a brief that suits them and then tend to stick to it. Obviously flexibility should be maintained to accommodate the experience of the client and the amount of information that the client has provided, but in general the approach is in reality fairly consistent. The comments given opposite, made during the case study interviews, are typical.

'We've had the same architect in two or three times and I know how he is going to take a brief, I know what he is going to come up with before he does.'
(Architect turned project manager, on long term contract with a large client)

'How do I know that I have got sufficient briefing information? Good question, because I know what the design team need, I know when there is enough information. I suppose this is something that I do not even think about.'
(Surveyor, client's internal project manager)

These comments demonstrate that experience appears to be the major driving force behind 'brief-taking' (a dangerous expression?) as far as construction professionals are concerned. Experienced brief-takers will have developed, over the years, their own set of internalised briefing 'rules' to which they continually refer. Even though a number of published briefing guides exist, the case studies drew

very few brief-takers made any real use of [published briefing guides]

attention to the fact that very few brief-takers made any real use of them and preferred to rely on their own experience. Such reference material was considered to be too prescriptive or vague and the following comment made by an architect was typical: '*a briefing guide sounds like a load of bull*'.

In addition, there was little evidence to suggest that brief-takers consulted with other professionals in order to learn how other people approached briefing. Finally, very few construction professionals appear to have actually received any worthwhile education relating directly to briefing during their original training and so have very little option but to rely on their own experiences.

These points all raise questions about the effectiveness of a construction professional's favoured briefing approach. If professionals rely mainly on experience, how do they know when there are gaps in their knowledge or whether some of their standard rules are no longer applicable?

Insights from work on human error

Reason, J. (1990) *Human Error*, Cambridge University Press, Cambridge.

A useful perspective is provided by human error theory. Reason (1990) has carried out a wide-ranging synthesis of human error mechanisms across many industries and situations. This work often relates to disaster situations where the consequences are patently tragic, but similar mechanisms seem to underpin many of the problems in briefing. The disasters may not be so newsworthy, but they are none-the-less real and subtly affect people's lives for decades after the construction project has been completed. Reason has tried to categorise the reasons why things go wrong. He has suggested that rules that have previously worked in the past may continue to be utilised by people even when they are no longer appropriate; he terms these *strong-but-now-wrong* rules.

strong-but-now-wrong rules

He proposes that there are several factors that may be responsible for the misapplication of previously successful rules. Some of these could readily be applied to briefing and could help to explain why brief-takers are often unwilling to part with their tried and tested methods. Not all briefing failures, however, can be attributed to the misapplication of a certain rule or method. On occasions failures may simply be the result of a lack of knowledge on the part of the brief-

failures may simply be the result of a lack of knowledge on the part of the brief-taker

taker. Table 1.2 illustrates both rule-based and knowledge-based factors that may affect briefing, using Reason's (pp. 74–96) categories as a framework.

This focused analysis contributes some interesting insights on the problems faced by construction professionals when involved with briefing and may help to explain why tried and tested methods of briefing are sometimes used inappropriately. It is worth mentioning that even though the analysis relates to the construction side, similar failures have also been identified on the client side. For example, experienced clients may want a new building that draws on similar projects with which they have previously been involved. However, a construction professional may be able to point out that there may be a better way of approaching things and that there is often more than one solution.

Insights from decision-making work

Another useful perspective on the developing picture of briefing is that of the debate over rationality in decision-making. Much of the previous work on briefing has assumed a model of rational decision-making. Such work begins with the client, assisted by the designer, setting out a pre-existent and consistent set of goals in the brief. It is then the task of the designer to make a series of rational choices about the form of the building that will enable these goals to be realized. The reality emerging from the case studies is very different and rather messy. Table 1.1 revealed a jumble of conflicting and confused aims, insufficient or over-detailed information and a lack of clear responsibilities. Despite apparently badly flawed processes, most of these projects reached passable, and in some cases excellent, conclusions. A theory of rational decision-making would have predicted disaster.

jumble of conflicting and confused aims

One possible explanation for the emergence of order from chaos in briefing may come from the critique of rational decision-making developed by March and Olsen (1997). Goals, they argue, are not fixed but in a state of flux. Organisations may need to be able to question their goals in order to develop. But, in order to carry out this questioning, an organisation needs to suspend rationalistic mentality and

March, J.G. and Olsen, J.P. (1997) The Technology of Foolishness. In *Organization Theory*, (ed. D.S. Pugh), 4th edn, pp. 339–352, Penguin, London.

Table 1.2
Understanding the data from a human error perspective

REASON'S FACTORS	EXAMPLES FROM CASE STUDIES
Rule based failures	
Rule strength If a particular approach has been used often in the past with success, there is an increased chance that it will be used again.	Every time a particular briefing method is used with success, it reinforces the notion that a particular approach is the best way of eliciting a client's brief. Unless major problems arise as a consequence of a specific method, then it is extremely unlikely that a brief-taker will ever change their style or even consider a different one. In many cases the possibilities for changes are further prohibited by the short timescales often involved in briefing; if a brief-taker only has a couple of weeks to find out all the relevant information, then not unsurprisingly a tried and tested information-gathering method is activated.
Rigidity If a rule has been applied successfully in the past there is an overwhelming tendency to apply it again even though a simpler solution may be more appropriate.	This may be the reason behind the lip service category. Construction professionals sometimes appear to want to try new ideas, but the internalised rules are so strong that they always win and people revert back to their old ways of collecting briefing information.
Redundancy Repeated encounters with a similar problem configuration allow the experienced trouble-shooter to identify certain sequences that tend to co-occur. The problem solver learns that truly diagnostic information is contained in certain key signs, the remainder being redundant.	Similarly an experienced brief-taker may encounter certain problems time and time again, particularly if dealing with the same building type or even the same client. This familiarity means that the brief-taker learns that certain key cues generally tend to be more important than others. This may be why architects are often criticised for beginning to design too early; having identified a familiar pattern they feel that they have enough information to begin drawing. Thus information which may be equally important as far as the client is concerned may not receive as much attention as they think it should.
Information overload Information overload may occur when a problem solver is dealing with a complex situation and so it may be difficult to identify exceptions to the traditionally implemented rules.	Briefing often produces a vast amount of information, often contradictory and sometimes irrelevant. Most people find it difficult to process such complex information, so the person trying to make sense of such information may focus on just a few things, often those that appear familiar, and so may miss exceptions to the traditional rules that could actually be important. This may help to explain why, even when different data collection methods have been employed, the final design may fail to incorporate some of the findings.

Continued

Table 1.2
Continued

REASON'S FACTORS	EXAMPLES FROM CASE STUDIES
Knowledge based failures	
Selectivity Not knowing which are the important factors, so attention could be given to the wrong features or not given to the right ones.	If only one specialist is responsible for taking the brief they may focus on asking questions about their area of expertise. For example architects may give priority to aesthetic or space issues rather than cost or M + E when these may be the most important aspects. Thus these factors may only be given proper consideration at a later stage when other consultants are brought in.
Out of sight, out of mind Undue weight may be given to facts that come readily to mind, while those facts that are not immediately obvious may be ignored.	The sheer amount of information produced during briefing means that facts may be overlooked, especially if a brief-taker is dealing with a building type that they are unfamiliar with. Unless the client participates fully, certain facts may remain undiscovered.
Confirmation bias Once a hypothesis has been formed, even if it is based on early, impoverished data, a problem solver may be loath to part with it even if contradictory evidence comes to light.	This may be particularly true when an architect is responsible for taking the brief and designing the building. If architects began to design before collecting all the relevant information, they may be reluctant to change their schemes at a later stage.
Overconfidence Problem solvers are likely to be overconfident in evaluating the correctness of their knowledge. They will tend to justify their chosen course of action by focusing on evidence that favours it and by disregarding contradictory signs.	Experienced brief-takers especially may assume that they have covered all the options and that their assumptions are correct. In addition, clients may rely too heavily on construction professionals' abilities to interpret their requirements, hence few checks are made by the clients and incorrect assumptions are allowed to dominate without being challenged.

allow the free play of intuition. March calls this suspension of rationality the 'technology of foolishness'.

Perhaps the reality of briefing, as opposed to its rationalistic explanation, contains elements of the technology of foolishness. An organisation that requires a building is in a state of change. It will no longer be quite the same after the new building is complete. In reality the organisation does not set out a clear set of objectives for the new building. No initial brief is exhaustive or immune to alteration. As the project develops the existing goals are challenged, sometimes abandoned and often developed. The result is the observed

An organisation that requires a building is in a state of change. It will no longer be quite the same after the new building is complete

anarchy. Eventually the guiding principles for the new building emerge and the project usually reaches a satisfactory, but not optimal, conclusion. The goals do not precede the action but rather emerge from it.

If briefing does not conform to a rational model then there are important implications for any attempt to improve practice. Formal process models, comprehensive checklists and the like assume pre-existing goals and may stifle the play of intuition. Innovative techniques should, on the other hand, seek to encourage the development of new goals and leaps of the imagination. There is, at the minimum, a tension here!

Better briefing

There is little evidence that 'good practice' *per se* actually has the expected effect when used in practice. On the contrary, examples abound of good practice not working, or being undermined by lack of commitment (lip service). There are also examples of 'bad practice' working well in practice. The implication is that to achieve successful briefing something more than traditional best practice ideas are needed.

to achieve successful briefing something more than traditional best practice ideas are needed

Much of the best practice advice provided over the last thirty years has been based on a purely rational perspective of the construction process. It has in effect said:

- If only people would plan carefully
- If only people would stick to their decisions
- If only people would do what they said they would do
- If only people wouldn't look out for their own short-term interests
- If only people didn't seek and abuse power.

This amounts to saying '*if only people were not involved everything would run smoothly*'! But, people are crucially involved and efforts at better briefing must accept, accommodate and work with this reality.

people are crucially involved

The pure rationality of the briefing process is also compromised by the fluid nature of the parameters. This is implicit in the nature of the operation. Clients are buying something they cannot see until it is finished, and very often the building produced will make a significant difference to how they will carry out their work. Thus, there can easily be a situation where the client finds it hard to imagine how they will operate within the building, as well as what the building will actually be like in physical reality. The situation is exacerbated by the fact that client organisations often seem to 'manage the margin'. In fact this is logical for them and amounts to management by exception. So, a client will be very clear about employing an extra member of staff, but can find it very difficult to tell an architect how many staff they want to accommodate in global terms. In one project the client stated that they had somewhere between 2500 and 3500 staff! But this lack of certainty had only become an issue because new accommodation was to be built for them.

client organisations often seem to 'manage the margin'

The difficulty for clients of imagining a new organisation in new accommodation is really quite wide-spread. It can be seen in its extreme form where the client organisation is actually using the construction project to re-engineer their organisation by physically re-grouping people, breaking down old structures and facilitating new flexible ways of operating, say through shared spaces or hot desks. In this situation the goals are being politically negotiated all the time, but even in a normal project there is a transition from a vague notion to a tangible artefact. The briefing process must support the client through this journey from uncertainty to certainty in such a way that aspiration is turned into delight.

this journey from uncertainty to certainty

On the construction industry side there is the contrasting danger that they have seen it all before. They have experience. Whether it is relevant and helpful is a moot point and will vary tremendously from practice to practice, individual to individual and project to project. However, in general it seems that experience can act as a barrier to the acceptance of good practice proposals.

experience can act as a barrier

Overall, best practice advice to date appears to have had little impact. There is the difficulty in persuading people to shift from their experience-based practices, especially if the alternatives proposed are detached from reality. The reality includes reflecting the inherent non-rational turbulence of the construction process and the importance of the human dimension.

The remainder of this book seeks to address the issues in a way that enables those involved in briefing to take up the suggestions and link them to their own experience. It does not seem far-fetched to suggest that, in fact, all the players are *trying* to act rationally, but within the context of their roles in the process and their particular perspectives. To move from this position, flexibility and breadth of vision are needed. Thus, the emphasis of what follows is not on how to do good briefing *per se* but rather how to see the way forward to achieving progressively *better* briefing.

how to see the way forward to achieving progressively better briefing

Key improvement areas

This section focuses on the identification of five key improvement areas for better briefing and is based on extensive analysis of a large number of projects (sixteen projects were studied in detail). As the view that emerges is quite simple, details of the extensive work that underpin the view set out here are given in Appendix C.

An approach was taken that dealt with the detailed data at project level from many different perspectives. A lot of information was gathered and the major thrust was to move from identifying problems and their causes to proposing actions in key solution areas. These were actions which it appeared would have the greatest impact on the largest number of problems that had been identified. The technique used (which may well have utility in the briefing process itself) was to map links between context issues and briefing problems and propose solutions. This was carried out in the following eight areas that had been identified as particularly problematic at an interim stage of the analysis.

- Assessment of clients' needs
- Roles and responsibilities
- Programme
- Type of information
- Visualisation
- User involvement
- User groups
- Project constraints

Mapping from problems to solutions

At the end of the chapter the 'maps' for each of the eight areas are shown as Figures 1.4–1.11. These give a rich impression of the sorts of issues and connections identified and we would expect anyone involved in construction to recognise many of the individual links. The value of the maps is that the information is organised, both into the categories and running from context through problems to solutions.

For all the value of the maps in making the issues visible, they are in fact only an interim step to the identification of the key solution areas for better briefing. This follows below.

Key solution areas

If improvement is important to resolving briefing problems, then identifying key solution areas is an essential stage in the process. The maps in Figures 1.4–1.11 clarify the building blocks of the problems and identify potential solutions, but *key* solution areas span across the eight maps. Taking just the proposed solutions from the centre of each map, high-leverage solutions were sought. These are solutions that have a positive impact on a large number of problem areas. As a result of this analysis, five key solution areas were found.

high-leverage solutions were sought

Two major areas were identified in which action could address problems throughout the briefing process (see Figure 1.3(a)). The first and main area we have termed 'empowering the client'. This has to do with helping the client become an effective client in the project process and beyond. Linked to this is the area of 'managing the project dynamics' which is related to taking the brief right through the construction projects as a pervasive issue. Great progress could be achieved by addressing just these two areas, but three other supporting areas were also identified as shown in Figure 1.3(b). 'Appropriate user involvement' is an essential component of

Figure 1.3(a)
Major solution areas

Empowering the client
Managing the project dynamics
Appropriate user involvement
Appropriate visualisation techniques
Appropriate team building

Figure 1.3(b)
Major solution areas and three supporting issues

an empowered client in many situations. 'Appropriate team building' linked closely to the effective management of project dynamics. 'Appropriate visualisation techniques' underpinned (and often undermined) actions in all of the other areas.

five key solution areas

The five key solution areas have been identified and it is argued that action in these can significantly improve the effectiveness and efficiency of the briefing process, however it is currently carried out.

Each of the five areas will now be considered in more detail in a separate chapter. For each the principles involved will be discussed first, with illustrations drawn from relevant theory, then the main factors will be set out with detailed case study examples and commentary from a construction perspective.

At various places later in this book, references will be made to the five areas in marginal notes. To help you quickly see the connections, icons have been created for each of the five areas. These are set out below.

- ***Empowering the client***
  This icon depicts an energising lightning flash as the client is empowered to play their role.

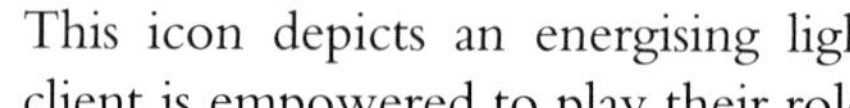

- ***Managing the project dynamics***
  This icon shows an 'S' curve typical of many projects, but within the main development thrust are many smaller activities to be done.

- ***Appropriate user involvement***
  This icon suggests the idea of a core team relating interactively with a wider community. 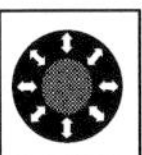
- ***Appropriate team building***
  This icon stresses the importance of people and the interaction necessary to build an effective team. 
- ***Appropriate visualisation techniques***
  This icon represents the idea of providing and using different views of the building in question and associated issues. 

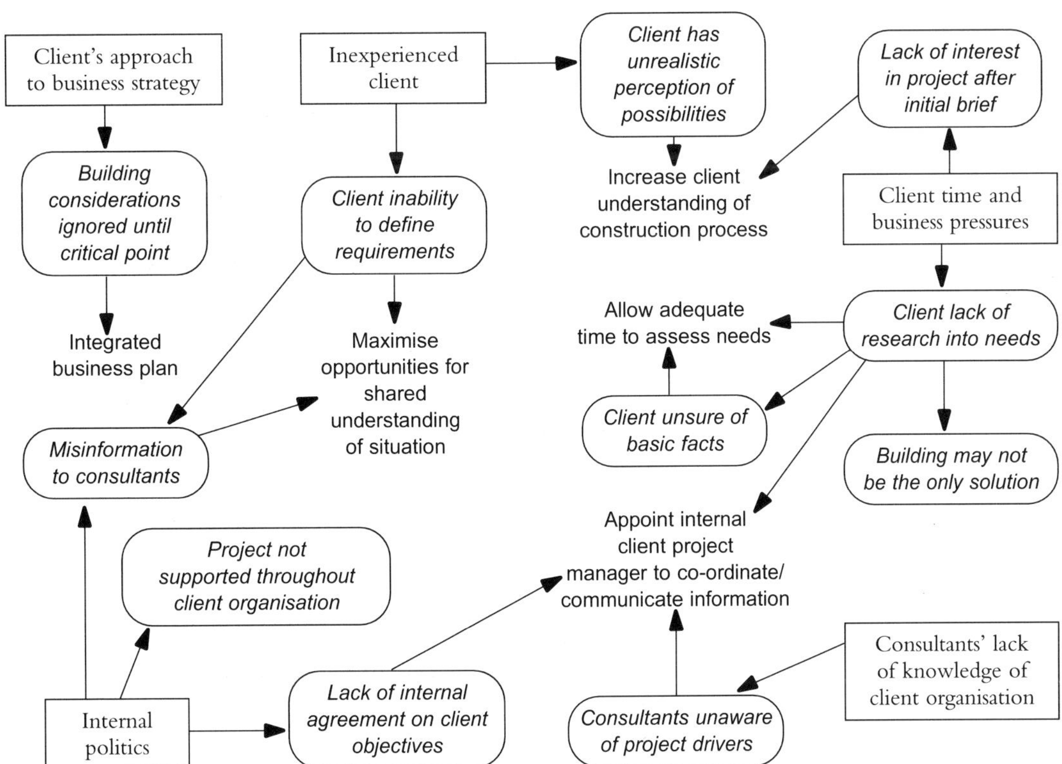

Figure 1.4
Assessment of client needs; ▭ = context; ▢ = problem; no border = solution

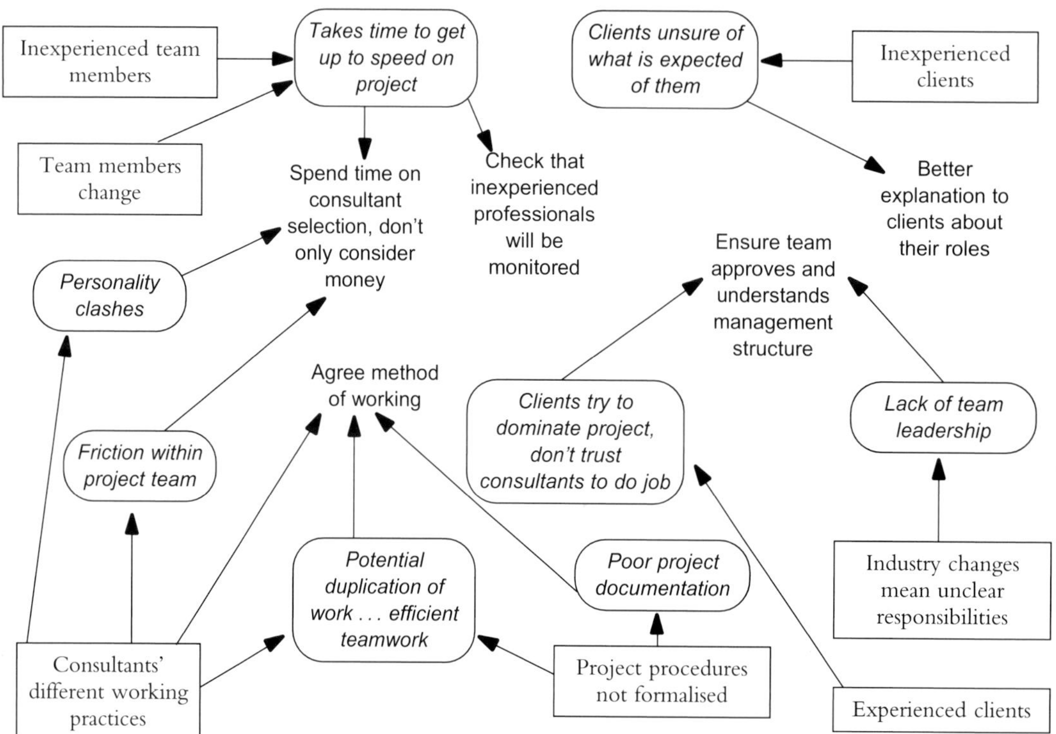

Figure 1.5
Roles and responsibilities; ▭ = context; ▢ = problem; no border = solution

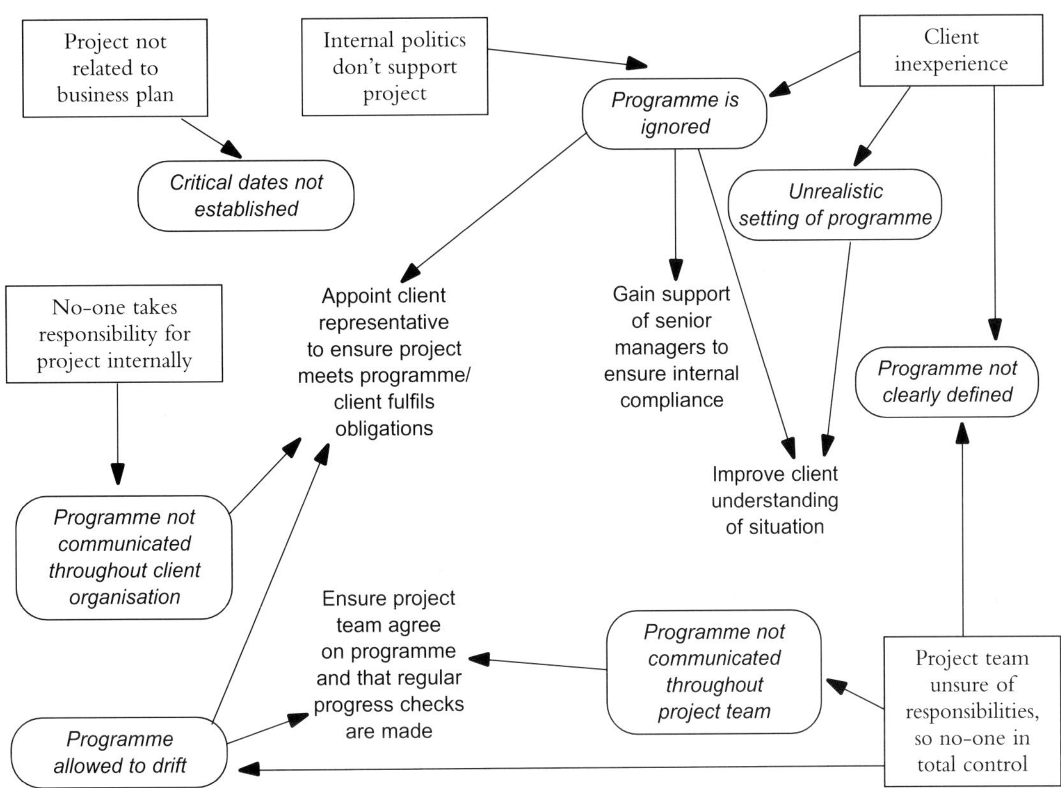

Figure 1.6
Programme; ▭= context; ⬭= problem; no border = solution

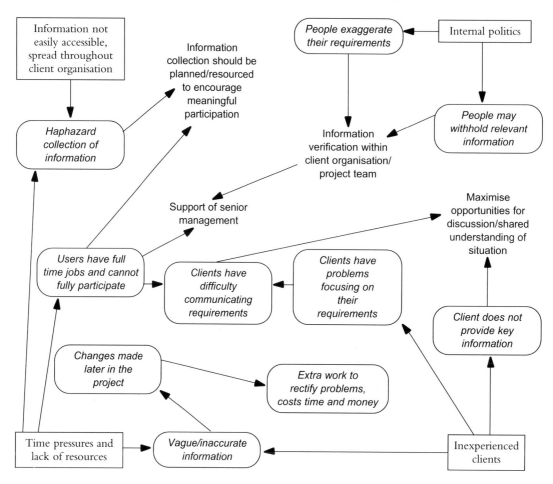

Figure 1.7
Transfer of information; ▭ = context; ⬭ = problem; no border = solution

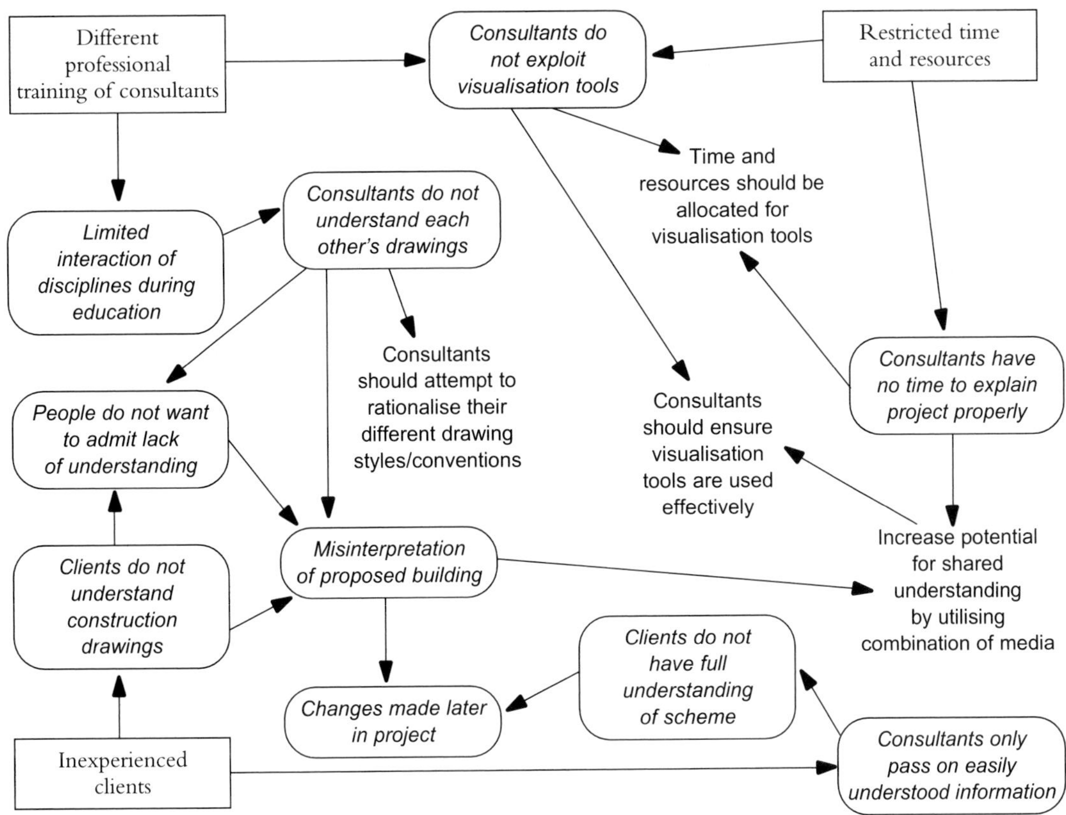

Figure 1.8
Visualisation; ▭ = context; ⬭ = problem; no border = solution

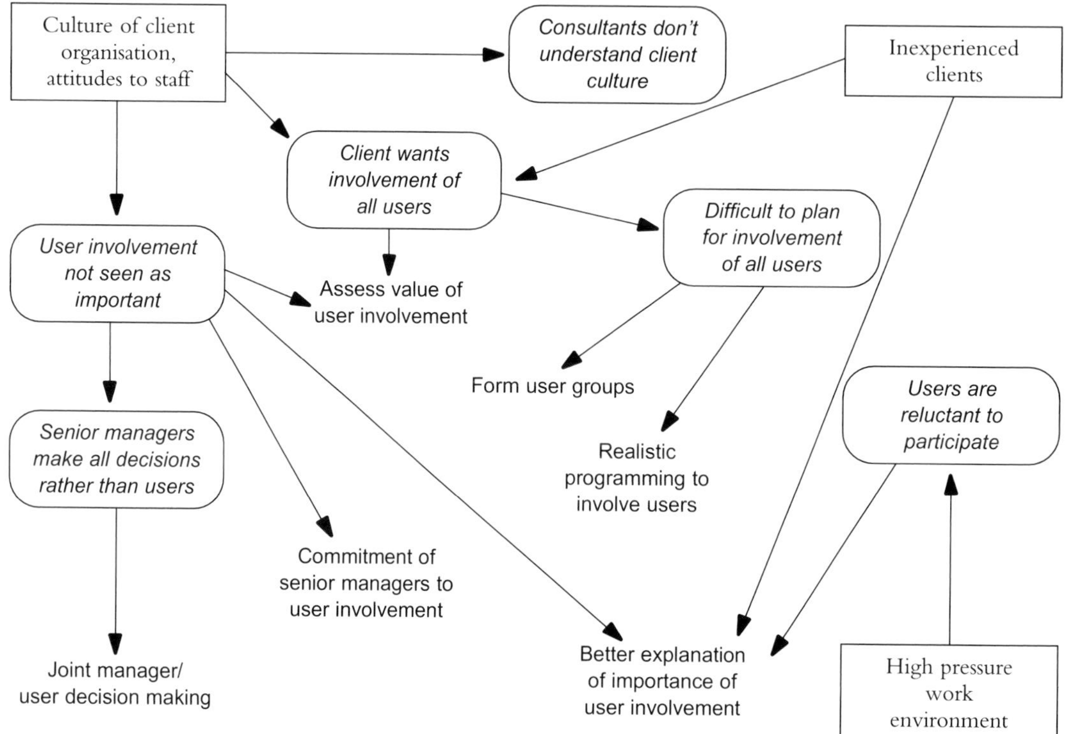

Figure 1.9
User involvement; ▭= context; ▢= problem; no border = solution

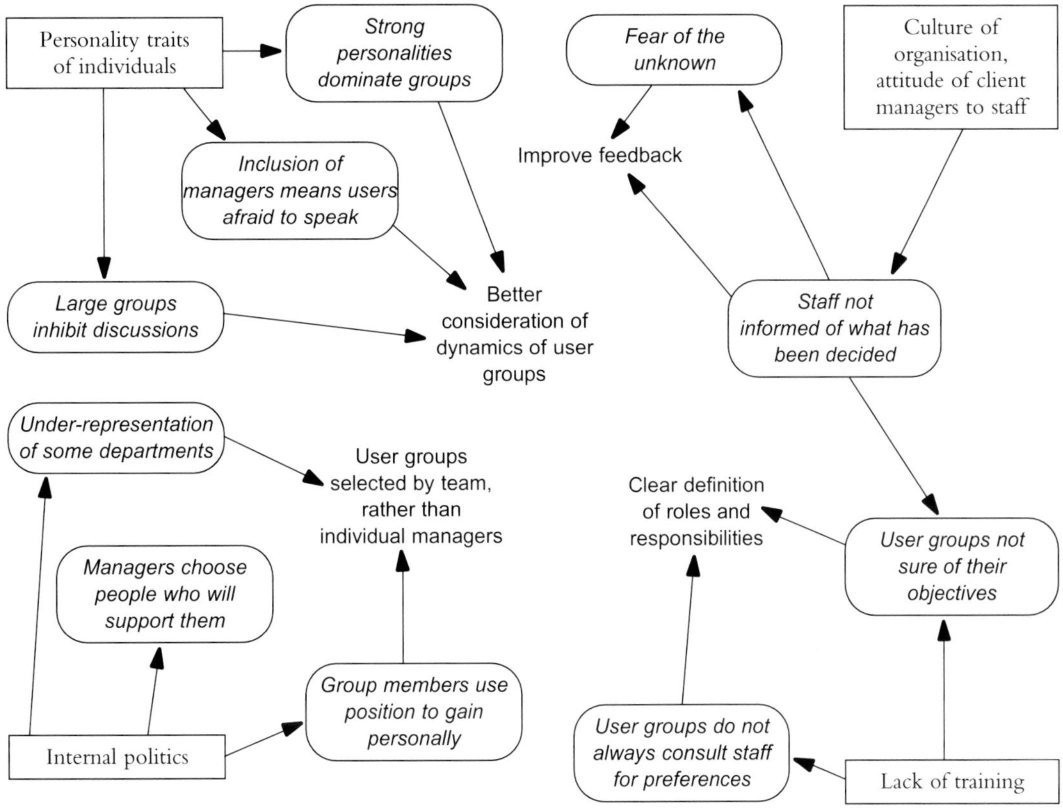

Personality traits of individuals

Strong personalities dominate groups

Fear of the unknown

Culture of organisation, attitude of client managers to staff

Inclusion of managers means users afraid to speak

Improve feedback

Large groups inhibit discussions

Better consideration of dynamics of user groups

Staff not informed of what has been decided

Under-representation of some departments

User groups selected by team, rather than individual managers

Clear definition of roles and responsibilities

User groups not sure of their objectives

Managers choose people who will support them

Group members use position to gain personally

User groups do not always consult staff for preferences

Internal politics

Lack of training

Figure 1.10
User groups; ▭ = context; ▢ = problem; no border = solution

Consultants' fees are squeezed

Consultants under constant pressure

Time pressures

Large number of competing demands requiring reconciliation

Utilisation of lower grade staff, may provide inaccurate advice

Thorough consultant selection process, obtain commitment of senior consultants

Need to make cost savings later in project

Establish programme, highlighting critical dates

Allow time early in project to identify client need

Increased research into/ prioritisation of constraints

Lack of in-depth assessment of different project constraints

Project team and client should identify realistic funding limits early on

Nobody is fully aware of constraints

Changes made later in project as more constraints discovered

Failure to recognise time pressures on project

Lack of communication throughout project team

Financing arrangements are unrealistic

Inexperienced client

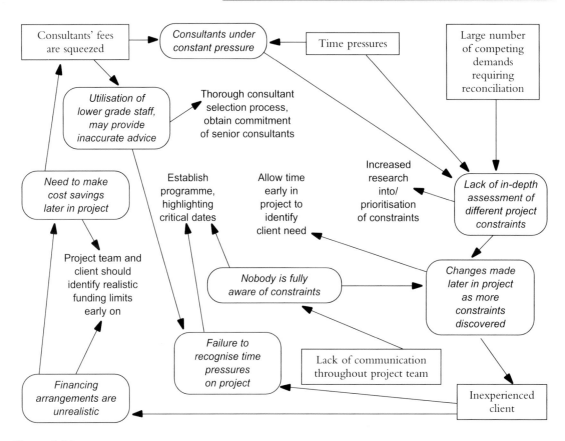

Figure 1.11
Constraints; ▭= context; ▢= problem; no border = solution

2
Empowering the client

2
Empowering the client

Discussion

It is extraordinary that whenever questions of briefing are discussed amongst construction professionals the first thing they do is blame the client. *Clients don't understand construction, they don't provide information on time and, worst of all, they change their minds!* An industry that blames its clients for the problems that exist in a key process such as briefing is doomed to failure. It will certainly not be able to improve the briefing process so long as it locates the reasons for the problems as being outside of its control. This is not to say that clients are not a 'problem' for briefing as it is currently carried out, but, being objective, this is more a criticism of the current processes than of clients.

the client may not always be right, but they are paying the bills

In principle the client may not always be right, but they are paying the bills. If they are having difficulty in providing timely and appropriate information for the construction process then it is incumbent upon those more knowledgeable in the construction industry to identify what help they need and to provide tailored assistance so that they can act effectively in the process.

a creative and interactive relationship . . . hinges on construction understanding clients

The fact that there is a construction project means that a client has identified a need for more, or different, space and this will be predicated on some need generated by their primary business. In this area the client is likely to be very expert and, at a minimum, why should they have to know any more? Provided they can put into words what they are trying to do in terms of their aspirations in their area of activity, then it is incumbent upon the construction industry to endeavour to understand and interpret it into a built solution. It may be that a much more creative and interactive relationship builds up, but it is not essentially dependent on how clients understand the minutia of construction. Rather it hinges on construction understanding clients.

In terms of their ability to effectively interact with the construction industry, clients will, however, have varying levels of confidence and competence to play an active role. It is useful here to draw upon leadership theory (Hersey, Blanchard and Johnson,1982; 1996). Work in the leadership area indicates a possibility of different styles of interaction. These styles are shown in Figure 2.1(a) and are generated by different combinations of task or people orientation. The key factor to draw from leadership theory is that there is no one right leadership style, rather it depends on the 'maturity' (confidence and competence) of the people you are dealing with. This additional factor is shown in Figure 2.1(b) and a

Hersey, P., Blanchard, K.H. and Johnson, D.E. (1996) *Management of Organisational Behaviour: Utilising Human Resources*, 7th edition, Prentice-Hall, New Jersey (4th edition, 1982).

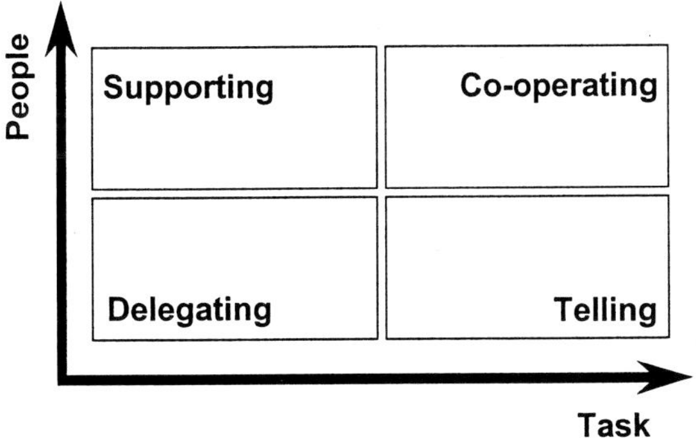

Figure 2.1(a)
Leadership styles. Based on Hersey and Blanchard, 1982

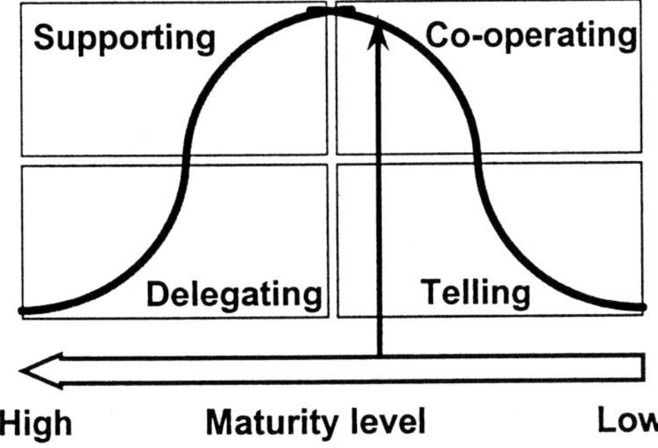

Figure 2.1(b)
Contingency leadership

bell-shaped interaction line has also been added. Thus, for someone of low maturity level a line taken up vertically to the bell-shaped curve shows that a 'telling' style would be the appropriate sort of relationship. At the other extreme for a high maturity person or group a 'delegating style' would be the appropriate relationship. In between 'co-operating' and 'supporting' styles are suggested. So, the style that should condition the relationship is driven by the maturity level of those you are dealing with. This in other words is a way of identifying and providing what the people you are dealing with lack.

Let us relate this specifically to briefing results in the representation given in Figure 2.2. The axes have been clarified as 'technical knowledge needed' and 'support needed', leading to four client types, typified with a brief statement. A client with very little knowledge of construction, but not really needing much support and encouragement would appear in the bottom right hand sector. This could be a small factory owner who simply wants some extra space and doesn't really care how it is achieved. Private individuals extending their own houses often differ from this in that they do need support as well as technical knowledge. What they are doing is very personal and the outcome matters to them a great deal. Thus an input from the construction professional both of technical knowledge and support is necessary to help them through it. A different situation is presented again by the client that has technical

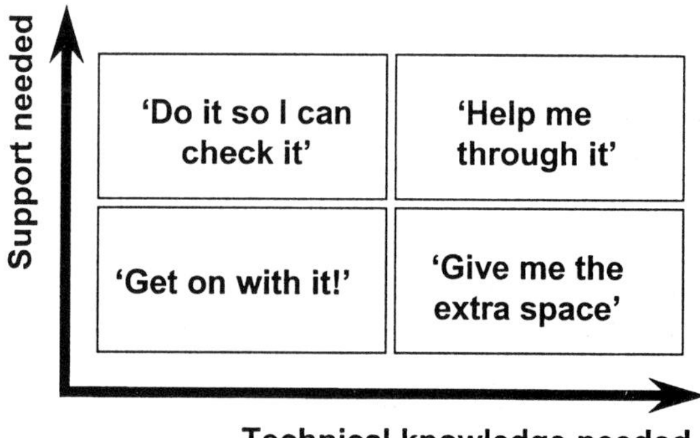

Figure 2.2
Contingency briefing

knowledge and is really simply buying additional labour for something that they could do themselves, if they had time. Put this into an organisational context where value for money has to be justified and the relationship revolves around doing the work so that the client can check it. A fourth alternative exists where the client is technically able and responsible only to themselves. Here the ground-rules may well be a rather unforgiving 'get on with it'. This could be where the client is a property developer.

clients differ considerably in their needs

The point which should come through the above examples is that clients differ considerably in their needs when dealing with construction and assessing the input they need for technical knowledge and support and encouragement is an important way of identifying how clients can be empowered in order to be effective in the briefing process.

Case studies

An analysis of the case studies indicated that 'Empowering the client' can be sub-divided further. A summary is shown in the list below. These are the main factors that clients should focus on with the support and guidance, where appropriate, of the construction industry.

- *Clients should be knowledgeable about their own organisations*
- *Clients should be aware of the project constraints*
- *Clients should understand the basics of the construction process*
- *Clients should understand their roles and responsibilities*
- *Clients should maintain participation in projects*
- *Clients should gain the support of senior managers for projects*
- *Clients should appoint internal project managers to manage projects*
- *Clients should integrate business strategy and building requirements.*

Clients should be knowledgeable about their own organisations

It should be remembered that clients, even those new to construction (naïve clients), are the 'experts' with regard to

'We were told that they had a brief, but when it came to it they had no such thing, they had a whole series of conflicting aspirations among the various people who were involved.'
(Architect)

'We ask them if they can write down how the department works and how it interacts with others. You find that it is usually a mess and they just don't know. So we have to challenge them and get them to think about it. Then again, quite often we go into these companies and we don't understand the terminology, it's a foreign language.'
(Architect)

'I think a lot of architects are very arrogant about what they think they know. Our philosophy is we do not know it all. We listen, we haven't got a set pattern, we tailor the answer to the client. Some clients say "Why do we need you, we give you the answers and you just draw them up!" That is actually a great accolade, because if I've teased out what they really wanted and converted it into a piece of architecture they like, then I'm doing my job.'
(Architect)

their own organisations and how they operate. Even if the consultants have worked on many similar buildings before, they may well overlook a critical piece of information if it is not drawn to their attention. Each organisation is unique and what is important to one organisation may not matter to another. It is up to the client to impart as much relevant information as possible.

Clients, particularly naïve ones, may find it difficult to describe their operations to another party, hence consultants should try to become skilled in the art of questioning. They are there to interpret a client's needs, not define what those needs are. During the case studies some clients complained that they felt they were getting a standard solution, not one that had evolved from their requirements.

Most case study clients produced some sort of briefing document outlining their requirements. Very often these were seen as inadequate by the consultants as they did not provide the right sort of information and had obviously been prepared without consultation or agreement within the client organisation. Once consultants began to question clients, conflicting requirements often became apparent. It is probably impossible for clients to set down all their requirements in a briefing document and in-depth discussions will need to occur in order for the consultants to produce a truly suitable design. However, clients may find it useful to refer to the proforma in Appendix B. This outlines what sort of questions clients should be asking themselves when they are preparing a brief or discussing their requirements with consultants.

Good practice

One of the case study clients employed specialist consultants to study various aspects of their operations, such as room usage, work processes and circulation patterns. The consultant highlighted how the current building was being utilised and how/where people interacted with each other. This information suggested that the majority of people worked in isolation and did not visit each other's offices, but that when they met by the vending machine they exchanged ideas. This inspired the project team to create a design with a number of informal meeting areas which encouraged the transfer of information.

Clients should be aware of the project constraints

When clients, particularly inexperienced ones, come to brief the rest of the design team they are not always sure of their requirements. During the case studies there were a number of examples where the clients had not even worked out basic facts, such as how many people would occupy the proposed buildings or how much money was available for a project. If clients are unable to convey such rudimentary knowledge, consultants will be unable to provide appropriate solutions.

'There was a lot of uncertainty about who was actually going to move into the new building. Whenever we asked them (the client) who was going on which floor they never knew exactly. They always said they would tell us next week.'
(Architect)

When considering project constraints, clients should try to prioritise their requirements. In one of the case study projects, the client wanted the building to be completed by a specific date and so the project team designed a solution based around a steel frame that could be constructed relatively quickly. However, when the steel option was compared to a concrete alternative, the concrete was substantially cheaper, but slower to construct. The client then had to decide which was more important, time or money.

Good practice

Some building types involve a certain amount of repetition, such as offices and hotel bedrooms. In such situations it makes sense for clients to implement a policy of standardisation across these repetitive spaces, i.e. all offices should be limited to a specific size. This means that they can concentrate their efforts on the more specialist or important areas of the building. For example, one organisation limits workstations in open-plan areas to a maximum of 100 sq. ft. and partners' offices to 125 sq. ft. This approach means that the briefing process can be speeded up as the users' options are limited.

'We have various corporate standards dictated, space for secretaries, space for different levels of management. It is in our interest to reinforce the corporate standards because it makes our lives easier. Before the company issued standards it used to be a free for all. The heads of these departments are powerful people, if one of them wanted a large office, even my boss would have a problem arguing it out with them.'
(Client's internal project manager)

Clients should understand the basics of the construction process

'For a lot of our clients, it is the first time in their lives that they have been involved in a building project. We found that they didn't even understand the language we were using. We now have a completely documented project process. We write down what stages we will go through, etc. We then send it to the client. It is something that has become a routine part of our process now.'
(Project manager)

When inexperienced clients become involved with construction projects they are unlikely to understand exactly what is involved. Most clients will have heard of terms such as planning permission or building regulations, but will not understand in detail what these mean, or how they may impact upon the project. Terminology used by construction consultants on an everyday basis may also be confusing.

Good practice

Consultants should take the time to outline to the client the different stages of the project, the predicted timescales needed for each stage, alternative procurement methods, etc. Some of the case study consultants developed documentation specifically for this purpose, which could be adapted for each new client/project.

Clients should understand their roles and responsibilities

'There seems to be some confusion over roles. We take the lead with the consultants and tell them what to do, and we expect to be told what to do by the client. So we are carrying on, but we are not totally clear that what we are doing is what anyone wants.'
(Architect)

Inexperienced clients in particular may not be sure what they are expected to do during the construction project. They may believe that as they are employing construction professionals their role as client will be limited to supplying initial information and then they can sit back and wait for the designers to come up with an appropriate solution. In the majority of projects, however, clients will be expected to undertake specific duties. Consultants should ensure that clients are fully briefed on their responsibilities. In several of the case studies the clients were unclear about their remit and this led to tension among the various parties.

When discussing client duties, such things as participation, decision making and project management should be covered. For example, clients, or a client representative, will normally be expected to attend project meetings, so that they are aware of project progress. Even though many decisions may be taken by the design team, the client will be expected to make non-technical related decisions, such as choosing which solution is the most appropriate for the way the organisation operates. If users are to be consulted it may be preferable for the client to manage and co-ordinate their involvement.

'I think there should be someone on the client team that knows about buildings and understands what the implications of things are. It's just that they don't seem to trust you when you say things. In our eyes we are their employees and we have their interests at heart, but I think they think that we are part of a conspiracy against them as part of the building industry.'
(Architect)

Clients should maintain participation in projects

Client participation varies from project to project. Some clients like to provide the initial brief and then let the consultants get on with it; they are the professionals after all. Other clients want to be involved in every decision. The latter may sometimes annoy the consultants as progress may be slowed down, but at least the client knows exactly what is going on. The former situation may well lead to problems – consultants need to know that they are pursuing a scheme that continues to meet the client's requirements and this can only be achieved by sustained interaction with the client. Clients possess much more detailed knowledge of their operations that consultants ever can, and so they are more likely to spot potential problems or alternative possibilities if they continue to participate in the project development after the briefing phase.

Good practice

Some of the case study consultants maintained client commitment and involvement in the project by ensuring that the client signed off the project at designated points, which were stipulated early on in the project. These consultants felt that these formal sign-offs were necessary as the client was forced to really consider what they were agreeing to. These consultants acknowledged that even after these sign-offs there was still potential for clients to change their minds, but the clients had also been well briefed upon the implications of changes and so this occurred less often.

'We explained where we were in the process and then asked if we could freeze the layout. Everyone was given the opportunity to speak out and there were a few arguments. Everyone wasn't happy, but the vast majority agreed and put their names to it. I get them to sign off bits of paper, I always do that!'
(Project manager)

Clients should gain the support of senior managers for projects

People within client organisations will not always agree what criteria a new building should meet. This is particularly

'Come August I put pen to paper, saying to all the executives, "If we are going to do this and you want it ready on time, you need to start thinking about the brief and the space issues now, it's essential that we begin." That kind of disappeared into thin air. The only comments I got back said, "Look we're just going to reorganise the way we do business, everyone is going to have to change the way they work, they are going to change who they will be working with and you are asking everyone to think about space!" Of course when I look at it that way, they're right.'
(Client)

'We had to make sure that the client representative knew who he was representing – was it himself, his own direct group or all the other groups? It is difficult because if you are doing an area with six laboratory sections in it, you could have one of those people nominated as the client's representative, you have to be very careful that you don't end up with one big laboratory and five little ones. Self-interest does creep into it. After you put it into a document and point it out to people, it is actually a bit more difficult then for them to pursue the self-interest bit too much.'
(Client)

problematic when the proposed project does not have the full backing of senior staff. Senior managers who disagree with the principles of the project may not provide information on time or may withhold relevant material. This can potentially affect the progress of the project. Organisations should try to obtain internal agreement on what the project should achieve, so that all senior managers are working to the same goal. If consensus cannot be reached, then the consultants need to be made aware of any relevant internal political difficulties and should discuss with the client how they should approach the situation. Consultants need to know what information they can reliably work with or where they may come up against hostility or a lack of co-operation.

Clients should appoint internal project managers to manage projects

Unless a project or client organisation is particularly small, it is unlikely that the people who have initiated the project will actually be involved with it on a day-to-day basis from start to finish. Most organisations, therefore, will need to appoint a client representative project manager who will be responsible for managing the client's part of the project. Senior managers need to consider their choice of client representative very carefully and ensure that the appropriate person is briefed fully on their responsibilities. Representatives need to understand that they are representing other people and so must draw attention to their concerns; representatives should not just put their own ideas forward.

Client representatives can perform a number of roles. They can co-ordinate the project within the client organisation ensuring that deadlines are met, collect user information, make decisions on behalf of the client, etc. It is very important that the different parties confirm which roles a client representative will undertake, particularly with regard to decision-making powers. Consultants need to be certain that any decisions taken by client representatives will be supported by senior management and not overruled at a later date. Similarly client representatives should know which decisions are outside their remit and should be passed to senior managers.

Organisations also need to consider how much time the client representative will allocate to the project. Many client representatives are expected to carry on with their usual work, as well as performing their project-related duties. Quite often this means that the client representative is under a lot of pressure and is not able to dedicate enough time or effort to either job. The organisation has to decide on the relevant importance of the new building and free up the client representative's time as necessary.

Good practice

On some projects it may make sense for the client to appoint more than one internal representative, maybe from different departments within the organisation. This allows the workload to be divided. More importantly it means that issues can be viewed from a wider perspective and so decisions may be made without reference to senior management. But, a coordinated view must still be provided.

'They established an in-house project team to collect the user information. They decided who they needed to speak to from which department and when. Then they would call me in when they had something concrete to discuss.'
(Project manager)

Clients should integrate business strategy and building requirements

Consultants often complain that clients want buildings to be completed as quickly as possible in order to meet their business goals. This puts pressure on the consultants from day one and often means that the completed building is less than satisfactory, as all phases of the construction process, including briefing, are reduced to a minimum. In many instances the organisations concerned have failed to integrate their business strategies with the associated building requirements. New products or sections of the business have been developed leading to an increased demand for space, which the organisation needs to meet as soon as possible. However, organisations need to bear in mind that the lead time associated with construction projects means that they need to begin work on new buildings at least two years, normally more, before the space is actually needed.

'We had a five year development plan and it had been recognised that we would have to do something, but we weren't quite sure what the specifics would be.'
(Client)

Before clients procure totally new buildings they should take the time to determine whether this is indeed the most appropriate solution. Organisations should assess their current operations and accommodation to see if changes can be achieved within existing limits. A reworking of practices or the internal arrangement of a building may actually achieve the same goal.

Summary and self-assessment

Overall this chapter has stressed the importance of the client being empowered, either through their own efforts or with support from the construction professionals they are working with.

Table 2.1 focuses on the main factors necessary to achieve this end and provides a self-assessment tool for the reader to assess where effort is most urgently needed. The tool can be used for the reader's organisation in general, some part of it, or it could be used to assess a specific project and just those people involved.

The outcome of the analysis is the identification of those factors which are a high priority, either because they are not already met in practice or because little effort is currently being made to satisfy them, or, most powerfully, because of both. A table like this is provided at the end of each of the next four chapters. As this is the first time it appears Table 2.2 illustrates, with a hypothetical example, how it looks when it has been completed.

This information can be used in various ways. It can be the basis of an initiative to pick up the highest priority factor, but this would not make sense unless this corresponded to an urgent or important issue for the firm. So it may be better to take an important, but low priority factor. Another alternative is to design an initiative that will have a positive effect on more than one factor. In this way a high priority factor may be linked to an urgent issue for the firm.

It should be remembered that the factors highlighted at this stage are in only one area, albeit a very important one. There are four more, which together make up the 'five-box' model. It may be worth waiting until similar analyses have been done for the other areas and a combined summary created as shown in Chapter 12. It may then be even easier to link together high priority factors with issues that are important and urgent for the firm.

Table 2.1
Summary assessment of empowering the client

Instructions
For each of the factors listed below, circle one number in each of the two scales provided to indicate the extent to which the factor is currently met and the effort you are currently making to address the factor. A key for each scale is provided below.

Met?
This scale is to assess the extent to which each factor is currently met:

1 = Not met at all
2 = Met to only a slight degree
3 = Met quite well
4 = Fully met

Effort?
This scale is to assess the effort you are currently making to address each factor:

1 = No effort at all
2 = A little effort
3 = Quite a lot of effort
4 = Considerable effort

Factors	**Met?**	**Effort?**
E1. Clients should be knowledgeable about their own organisations	1 2 3 4	1 2 3 4
E2. Clients should be aware of the project constraints	1 2 3 4	1 2 3 4
E3. Clients should understand the basics of the construction process	1 2 3 4	1 2 3 4
E4. Clients should understand their roles and responsibilities	1 2 3 4	1 2 3 4
E5. Clients should maintain participation in projects	1 2 3 4	1 2 3 4
E6. Clients should gain the support of senior managers for projects	1 2 3 4	1 2 3 4
E7. Clients should appoint internal project managers to manage projects	1 2 3 4	1 2 3 4
E8. Clients should integrate business strategy and building requirements	1 2 3 4	1 2 3 4

Instructions (continued)
When you have done this for each of the factors, use the two scores to locate each factor, using its list number, in the grid provided to the right of this box.

Entries that fall in the black areas can be ignored. For the remainder, the lighter the area the more urgent it is for there to be a concerted effort on the factor concerned. In this way high priority factors can be identified.

These high priority factors can be combined with similar results for the other 'five-box' areas in Table 12.1 on page 133.

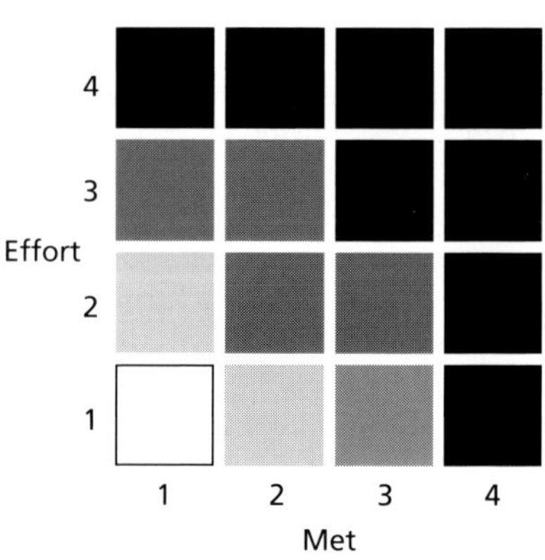

Summary assessment grid

Table 2.2
Completed summary assessment of empowering the client – Example

Instructions
For each of the factors listed below, circle one number in each of the two scales provided to indicate the extent to which the factor is currently met and the effort you are currently making to address the factor. A key for each scale is provided below.

Met?
This scale is to assess the extent to which each factor is currently met:

1 = Not met at all
2 = Met to only a slight degree
3 = Met quite well
4 = Fully met

Effort?
This scale is to assess the effort you are currently making to address each factor:

1 = No effort at all
2 = A little effort
3 = Quite a lot of effort
4 = Considerable effort

Factors	Met?	Effort?
E1. Clients should be knowledgeable about their own organisations	1 ② 3 4	① 2 3 4
E2. Clients should be aware of the project constraints	1 2 ③ 4	1 2 ③ 4
E3. Clients should understand the basics of the construction process	① 2 3 4	① 2 3 4
E4. Clients should understand their roles and responsibilities	1 ② 3 4	1 2 ③ 4
E5. Clients should maintain participation in projects	① 2 3 4	① 2 3 4
E6. Clients should gain the support of senior managers for projects	1 2 3 ④	1 ② 3 4
E7. Clients should appoint internal project managers to manage projects	1 2 3 ④	① 2 3 4
E8. Clients should integrate business strategy and building requirements	1 ② 3 4	1 ② 3 4

Instructions (continued)
When you have done this for each of the factors, use the two scores to locate each factor, using its list number, in the grid provided to the right of this box.

Entries that fall in the black areas can be ignored. For the remainder, the lighter the area the more urgent it is for there to be a concerted effort on the factor concerned. In this way high priority factors can be identified.

These high priority factors can be combined with similar results for the other 'five-box' areas in Table 12.1 on page 133.

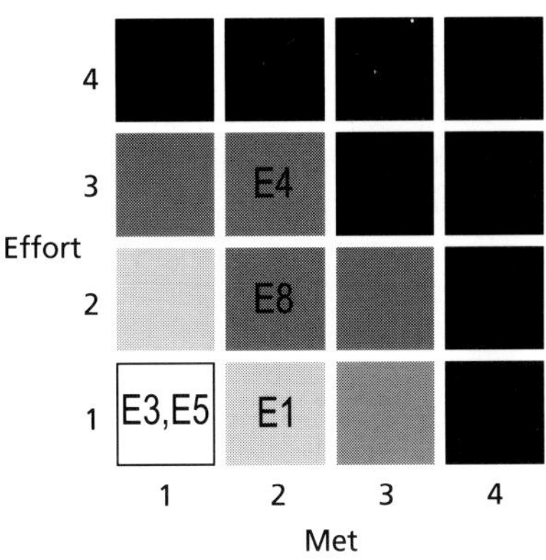

Summary assessment grid

3
Managing the project dynamics

3
Managing the project dynamics

Discussion

The importance of this area can be clearly illustrated by referring back to the discussion in Chapter 1 on the different conceptions people have of the brief. The 'brief' that is only taken off the shelf so that it can be clearly demonstrated to the client that the architect has exceeded the parameters that the client set led to the view of the brief as a 'cage' for the client. It is constructed as quickly as possible at the start of the project and the client placed in it, the door slammed shut and bolted. It is a physical document ranging from a brief letter of appointment through to a large formal statement of needs. In either case it is a dead document.

The alternative view emphasises the verb rather than the noun

The alternative view emphasises the verb rather than the noun. The focus is on the briefing process, but even so this is seen as something that happens at the early stage of the project and is superseded by subsequent stages that build from it. This is the perspective taken in the RIBA Plan of Work, but the emphasis is still to 'fix' the brief as early as possible.

📖 RIBA (1967 and regularly updated) *Plan of Work*, RIBA, London.

It is obvious to anybody involved in construction that, in fact, a joint decision-making process is involved and so the question of how early in the process decisions should be made, or whether flexibility should be retained, is central to how the process is managed. It is interesting to look at a simple analogy to this process which highlights some important implications. The following is a serious demonstration, but can also be an amusing party game. It involves various shapes which represent items of information. The exercise is based on one given by Edward de Bono in a book on lateral thinking.

📖 de Bono, E. (1971) *Lateral Thinking for Management*, McGraw-Hill Book Company (UK) Ltd., 1971.

Two shapes are presented as shown in Figure 3.1(a) and a volunteer from the 'audience' is asked to combine them to form a single shape. Typically this is achieved as shown in Figure 3.1(b) but it could be that the shape in Figure 3.1(c) is formed. The audience is then asked which they prefer and discussion ensues until someone says, '*but how are we supposed to decide?*' At this point the game can be simplified by giving a clear performance criterion of forming the simplest shape possible. Clearly then the arrangement in Figure 3.1(b) is superior. At this stage an extra shape is introduced and this is inevitably added as shown in Figure 3.1(d). Then two further shapes are added as shown in Figure 3.1(e) and various solutions are offered by volunteers, such as those shown in Figures 3.1(f) and 3.1(g). The fact is that at this stage it is not easy to find a simple solution. However, at some point after pursuing various blind alleys, someone who has typically sat back and not taken part until that moment, someone who has given themselves the time and space to think, will enter centre stage, take the whole arrangement apart and re-form it as shown in Figure 3.1(h). This is a very simple shape, but one which was not immediately apparent to most people in the room.

What does this seemingly trivial experiment demonstrate? Two main things come out. The first is that, if it is not clear what the performance criteria are then it is very hard to

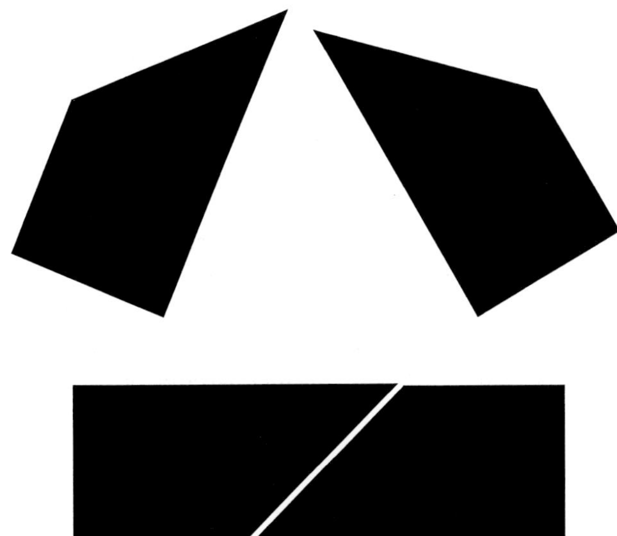

Figure 3.1(a)
de Bono's shapes experiment. Source: de Bono, 1971

Figure 3.1(b)
de Bono's shapes experiment. Source: de Bono, 1971

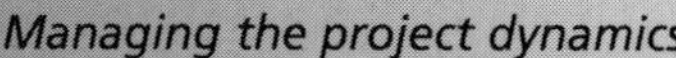

Figure 3.1(c)
de Bono's shapes experiment. Source: de Bono, 1971

Figure 3.1(d)
de Bono's shapes experiment. Source: de Bono, 1971

Figure 3.1(e)
de Bono's shapes experiment. Source: de Bono, 1971

Figure 3.1(f)
de Bono's shapes experiment. Source: de Bono, 1971

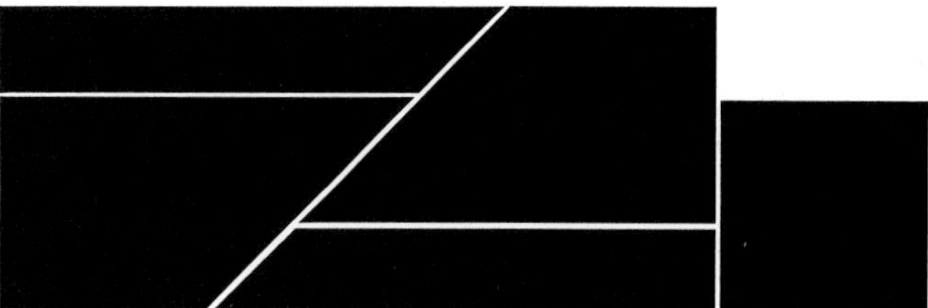

Figure 3.1(g)
de Bono's shapes experiment. Source: de Bono, 1971

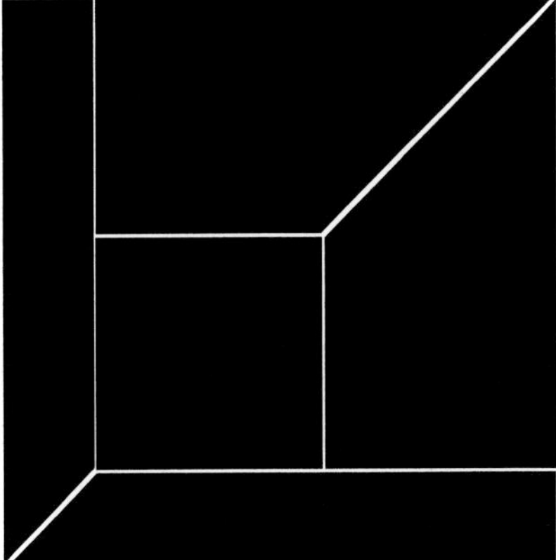

Figure 3.1(h)
de Bono's shapes experiment. Source: de Bono, 1971

choose between alternatives. In terms of briefing this means that diving straight in and talking about room sizes is almost certain to lead to confusion. First it is necessary to identify what the client is trying to achieve in *their* terms. This links back to empowering the client and majoring on the client's area of expertise, namely, what they do. The second thing that comes out clearly from the experiment is the impact of the order in which information arrives on the ability of individuals to identify optimal solutions. De Bono points out that there are in fact two decision-making features at work here: primacy and recency. Primacy is the overweighting of the first few items of information received and recency is the overweighting of the last items of information received. Arguably briefing suffers dreadfully from primacy in that the first few ideas that the client expresses tend to dominate the brief and may overshadow key inputs revealed in later discussions. Later in a construction process it is quite conceivable that recency takes over in that it is very common for specific problems on site to be resolved in isolation, with the associated danger of losing the overall coherence of the solution to the client's requirements. The antidote to the above dangers is also clearly shown in the game. Standing back and reflecting and allowing some time for the re-arrangement of the information at various stages can lead to new insights and optimal solutions that are not immediately apparent otherwise.

allowing some time for the re-arrangement of the information at various key stages

Spekkink, D. and Smits, F. (1993) *The Client's Brief: more than a questionnaire*, Stichting Bouwresearch, Rotterdam.

Spekkink, D. (1993) *Programma van Eisen (SBR 258)*, Stichting Bouwresearch, Rotterdam.

you decide as little as possible at each stage

Specifically in the area of briefing there is the interesting approach developed by the Stichting Bouwresearch (Building Research Board) in The Netherlands, which is depicted in Figure 1.1, and which seems to meet several of the conditions that have emerged above. Remember the underlying principle is that you decide *as little as possible* at each stage. The brief's development (termed programme development in the diagram) progresses in parallel with the development of technical solutions. Then at various key points the technical solutions are tested against the brief to the point that it has evolved, to ensure congruence. The work in The Netherlands has shown that there are quite a large number of features that do have to be decided quite early, but this is different to saying 'fix everything as quickly as you can'. In addition, the use of the brief as a live document running in parallel with the technical construction process is clearly something that can be carried right through the project to ensure that decisions made later still satisfy the overall objectives. It should also be evident from Figure 1.1 that the brief starts out as the client's aspirations in terms of a 'global brief' which is not expressed in construction terms, but instead focuses on functional needs.

There is a strong argument, set out above, for managing briefing as a process that extends throughout construction. In addition, developing the brief in parallel with, and separate from, the construction details, rather than leaving it behind, keeps the interaction between client and construction professionals alive. This is important, not only because clients' needs do change, but also because some creative solutions will only ever be revealed if the process supports close interaction between the client and the construction professionals.

Luft, J. and Ingham, H. (1955) The Johari Window, A Graphic Model of Interpersonal Awareness. In *Proceedings of the Western Training Laboratory in Group Development*, UCLA Extension Office, Los Angeles.

Bejder, E. (1991) From Client's Brief to End Use: the Pursuit of Quality. In *Practice Management: New Perspectives for the Construction Professional*, (eds P.S. Barrett and A.R. Males), pp. 193–203, E&FN Spon, London.

This is illustrated in Figure 3.2 in a diagram termed the Johari window. This was originally devised as a psychiatric counselling model by 'Joe and Harry' and so would appear to have great potential to address the ills of the construction industry! The diagram has been adapted to reflect the relationship between clients and their designers (Bejder, 1991). It shows that at the start of any relationship there is some shared knowledge that occupies the 'public' quarter. In addition, the client is excluded from a 'blind' area and the

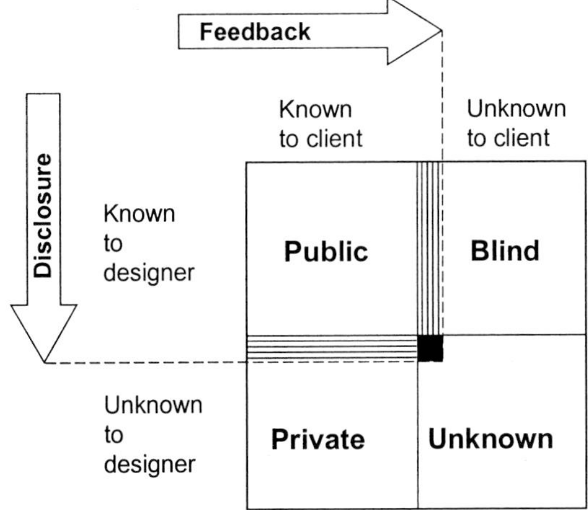

Figure 3.2
The Johari window. Source: Bejder, 1991

designer from a 'private' area. In addition there is an 'unknown' area in which information and ideas that neither party knows at the start are to be found.

To take a simple, domestic, example, the client may say to the designer that they want a kitchen extension to the back of the house and in so doing the client discloses information to the designer, so opening up the private area in Figure 3.2 that was previously hidden from the designer. The designer may then do some plans and come to the conclusion that an extension to the rear of the house is almost impossible to make work. As a result the designer may give feedback to the client that an extension to the side would be greatly preferable. This feedback opens up the blind area of the client who now better understands the situation they thought they knew pretty well anyway. The client then may disclose that they have in fact an option to buy land at the side of the house, therefore opening up more of the private part of the diagram and allowing a creative discussion to take place as to an alternative scheme with a kitchen extension to the side which neither party would have had in mind at the start of the process.

What has effectively happened is that the 'unknown' area has been revealed, but this would not have happened if the designer had simply taken the initial 'brief' and promptly placed the client in a cage. The importance of allowing time for feedback and disclosure to take place and for trust to

develop becomes all the more crucial when it is realised that the unknown quarter of the diagram is in fact infinitely big.

Minzberg, H. and Waters, J.A. (1985) Of strategies deliberate and emergent, *Strategic Management Journal*, **6**, 257–72.

A way of summarising this can be borrowed from work on the implementation of the strategies (Minzberg and Waters, 1985) as shown in Figure 3.3. The intended strategy is the initial idea, some of which turns out to be unrealisable. This leads to the deliberate strategy, which is adapted by emergent ideas, resulting finally in the realised strategy. The realised strategy is not in a very different direction from the intended strategy, but the whole process has been improved by allowing the unworkable aspects of the original ideas to be set to one side, while also allowing good ideas that arise during the process to be accommodated. The process of briefing should endeavour to display these major features throughout the project.

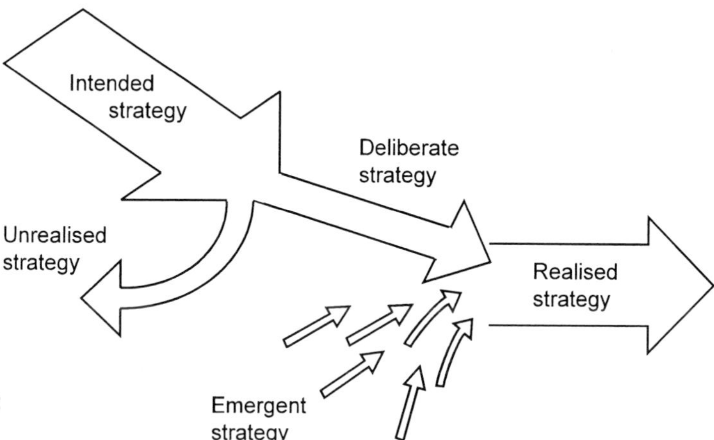

Figure 3.3
Strategy into practice. Source: Mintzberg and Waters, 1985

Case studies

When considering 'managing the project dynamics', project teams, that is clients and construction professionals together, should focus on the following factors.

- *Teams should establish project constraints at an early stage*
- *Teams should establish a programme, highlighting critical dates*
- *Teams should agree procedures and methods of working*
- *Teams should allow adequate time to assess client's needs*

- *Teams should validate information with the client organisation*
- *Teams should improve feedback to all parties throughout the project.*

Teams should establish project constraints at an early stage

All of the project team need to be sure that they understand, and agree, exactly what the budget will cover and how things such as inflation and VAT will be dealt with. In one of the case studies, the client and project manager had different opinions on how inflation would alter during the lifetime of the project. This led to friction amongst the team members as they were not sure if the budget for the building would be reduced if inflation increased at a greater rate than the client suggested. What they didn't know was that the client had kept some money in reserve that could be used if necessary, but that he did not want them to know this as he believed that the consultants would spend up to the maximum budget whatever it was.

'There had been a misunderstanding about the figure quoted to the architects. No-one was sure whether it had included VAT or not. The outcome was that the building suddenly had to shrink 4 metres.'
(Senior user)

Clients often need buildings to be complete by a specific date to coincide with their business plans. For example, a university will probably want its new accommodation blocks to be ready for the start of the new academic year. Unfortunately, such business pressures mean that clients often push for a project programme which is really unachievable as far as the rest of the project team are concerned. Clients who push for tight deadlines have to be prepared to ensure that their organisation fulfils its obligations, by taking decisions when necessary and by providing information on time.

'I think right from the start we had a misunderstanding about budget. They (the client) told us what budget was available, then progressively it has been cut back and cut back for reasons that we are unaware of. The reductions caused a lot of redesign work, although we weren't given any extra time for that. Whenever they lopped £1.5 million off the building budget they never extended the design period, they just expected us to take all that in our stride and shrink the building.'
(Architect)

Teams should establish a programme, highlighting critical dates

By the end of the briefing/design phase several of the case studies were already behind programme. This indicates that project teams often overestimate what can actually be achieved or do not build enough flexibility into the programme to allow for complications, such as a delay in obtaining planning approval. Additionally the programme is

'We started briefing in January and we were supposed to have a design freeze in August. Yes that famous design freeze! I mean it just didn't happen. If we had wanted to be awkward we could have refused to take

in any information beyond that point, but we carried on. We are still being briefed even though it's now January.' (Architect)

not always taken seriously by all relevant parties. If the project relies on input from users, clients need to make sure that their managers understand the programme, communicate its importance and allow time for users to consider their requirements.

Teams should agree procedures and methods of working

'The management contractor has produced a programme which shows when he wants the information and no-one has questioned it. However, one of the problems that has been happening recently is that these deadlines just pass by. We then discuss it at a meeting and they just say, "Oh it's going to be a week late." Then the project just drifts. A programme is there for a reason.' (Project manager)

In one of the projects a procedures manual had to be completed as part of the client's requirements, which would outline people's responsibilities and how the project would be managed. This manual was not really taken very seriously by the consultants (or client representatives) and was viewed as another piece of unnecessary bureaucracy. However, this project was approximately three months behind schedule by the end of the briefing phase as no-one was clear about what was supposed to be happening. The project would certainly have benefited if people had been forced to consider their roles and activities in greater detail.

Good practice

Project teams should produce a written *modus operandi* which outlines the management of the project and the responsibilities of each party. This needs to be communicated and commitment gained.

'I'm not sure what their role is anymore (client's project managers). They don't seem empowered to make any decisions, they only seem able to report back to one of the relevant committees who then supposedly make a decision. There is no timetable for them to make a decision, if they had a set period that would be fine, but they don't seem to. The situation could be improved by someone in the organisation having some authority, rather than every decision being a political mudbath.' (Project manager)

It is important that the consultants also understand how the client's procedures will relate to or affect the building project. For example, does the project have to be approved by a senior decision-making body at the end of one phase before it can proceed? If board approval is needed, what is the cycle for meetings? If the scheme is not approved, will the team have to wait until the next meeting before they can proceed with the scheme?

Requests for changes are an inevitable part of the building process. Client organisations do not stand still and often clients will want to incorporate changes that tie in with their business needs. Similarly, once users have thought about their requirements in more detail they may realise that they have missed something. During the case studies some users asked the consultants to incorporate changes directly without consulting the main client. These changes would often have resulted in an increase in floor space which the client would have had to pay for. Thus the project team should ensure that they establish specific procedures to deal with change requests.

Good practice

Several of the consultants had developed 'change forms' specifically to control how change requests were dealt with. These forms asked what change was required and why. Once the design team understood the reasoning behind a request they were often able to come up with a better, cheaper solution. More importantly, before the changes could be incorporated they had to be approved and signed-off by the client (or project manager, etc.).

'We set up change control on drawings. Once a drawing was frozen there were only two people who could unfreeze it or make changes.'
(Project manager)

Teams should allow adequate time to assess client's needs

It is probably impossible for clients to make decisions on all of their requirements at a very early stage in the project. Clients' original briefs are often minimal and it is only after they have had more time to think about things that their real requirements begin to crystallise. Consequently design teams should not rush to finalise the design as changes are extremely likely and schemes will probably have to be reworked. Architects in particular are keen to begin designing and many start sketching out schemes before they have amassed all of the relevant information.

'We issued a set of plans (to the client) that were going to form the basis of the tender. They were based on everything that we had been told so far. Then lo and behold I got a letter requesting ten pages worth of changes.'
(Architect)

Traditionally many clients have worked in conjunction with the architect (or design team) to produce a brief. However, the experienced clients studied believed that the design team were there to design, not work on the brief. These clients developed their own extremely detailed briefs as they knew which questions to address and what sort of information they needed to pass onto the consultants. This meant that the consultants didn't have to waste time collecting briefing information and could concentrate on designing the buildings.

'You have to go through the whole process of rationalising things; we are better placed to do that as the client's projects managers. The end users will always ask for twice as much as you can afford and twice as much as you can fit in the building. So we get it to roughly resemble what the business development plan will allow and then turn around and fire it at the design

team. They can then go off at speed and get the job done.'
(Client's project manager)

Good practice

Briefing and design stages could be handled by different people, that way all relevant information can be considered and the design team can concentrate on designing, rather than collecting information.

Teams should validate information with the client organisation

'By this stage we have 1:100 drawings of the layout of each individual department. These layouts are then checked by each of the departmental heads and the relevant user groups. If they agree with the scheme they will then sign it off. It is very important that they are signed off, otherwise you are not proceeding with an approved scheme.'
(Project manager)

'There was a very detailed brief set out of what this building had to do. But as the atrium has been developed we have lost space on two levels. It is a very interesting space, but it isn't a usable space and so we are missing some of the facilities that we had asked for.'
(Senior user)

Clients should be involved in the development of a project as much as possible to ensure that they are fully abreast of what is proposed. A well-informed client is less likely to request late changes. Some consultants insist that the client (and users) sign-off the project at specific points so that they know they are progressing with a scheme that has client approval.

Clients also need to be fully involved as projects progress as they are the only people who can really guarantee that their requirements are still being met. As projects develop everyone tends to focus on the details; one decision often has an impact on another and the client's original requirements are forgotten or misinterpreted. Project teams should check with the client that the developing scheme still satisfies the original objectives.

People tend to think of a brief as a written document which outlines the client's requirements; however, much briefing information is obtained through follow-up discussion. When clients outline their requirements verbally there are always opportunities for confusion as consultants will not necessarily interpret (or remember) a statement in the same way that a client does.

Good practice

'Whenever we speak to the client we always try to have more than one person there. If only one of us attends we are only getting one interpretation of what the client wants. With two people you get two views which I think is very valuable.'
(Architect)

One consultant organisation tries to make sure that two members of staff attend briefing meetings so that they can confirm with each other what was agreed or discussed with the client.

Clients often discuss their requirements with the lead consultant, often the architect, and this information is then passed onto the other consultants. However, the lead consultant will probably unwittingly sort or prioritise this information before passing it on, so that the other consultants will not always be aware of all relevant points. Additionally the lead consultant cannot be a specialist in all areas and so may not ask the client the sort of questions that the other consultants deem necessary.

Good practice

In appropriate circumstances it makes sense for complementary consultants to be involved in briefing discussions to ensure that the client's requirements are interpreted properly. For example, on a hospital project where building services obviously feature heavily, it may be a good idea to involve the services engineer in the briefing phase.

'In certain situations I find it beneficial to involve someone from a different discipline to work with me and the client on the brief. For example, in laboratory work the client will have a number of technical requirements and so I will work in parallel with a building services engineer as he can ask better technical questions than me.'
(Architect)

Teams should improve feedback to all parties throughout the project

As the project progresses members of the team should ensure that they are all working on the same version of the scheme. In some instances during the case studies, new information was acquired by one member and then only passed onto the consultant member that it really affected. Consequently decisions were taken which had knock-on effects elsewhere, but these only came to light at the next meeting. This caused friction between the members as drawings needed to be reworked in the light of this information.

The members of the project team are not the only people who need to be kept aware of project progress. During the case studies users often complained that once they had passed on their requirements they heard little more about the project until they occupied the new accommodation. If users are kept informed of progress they will be better prepared for the changes in their working practices that will almost certainly occur when they move into a new building.

'Through it all the key thing that we did was communicate (with the users). For example, we had workshops for the different representatives. People were cynical about these, but they were a very good way of communicating what we were up to. They then reported back to their departments – which they actually didn't do very well!'
(Project manager)

'We also developed a newsletter for the project. We sent these out to everyone who would be affected. We sent them out every so often and it told everyone why we were doing certain things and where we were up to. It worked an absolute treat.'
(Client's project manager)

Good practice

One of the case study client organisations produced a newsletter which was sent out to all involved users throughout the project. The newsletter described the progress of the project and the reasons for decisions. Consequently when the users finally moved to their new buildings they already felt familiar with the layout and so were able to work productively right from the start.

Summary and self-assessment

Overall this chapter has stressed the importance of the broad parameters of the project being clear and of the brief, and the interaction with the client, being kept alive throughout the project.

Table 3.1 focuses on the main factors necessary to achieve this end and provides a self-assessment tool for the reader to assess where effort is most urgently needed. The tool can be used for the reader's organisation in general, some part of it, or it could be used to assess a specific project and just those people involved.

The outcome of the analysis is the identification of those factors that are a high priority. An explanation of how the table is completed, and how the resulting information can be used, is given at the end of Chapter 2.

It should be remembered that the factors highlighted at this stage are in only one area. There are four others, which together make up the 'five-box' model. It may be worth waiting until similar analyses have been done for the other areas and a combined summary created as shown in Chapter 12. It may then be even easier to link together high priority factors with issues that are important and urgent for the firm.

Table 3.1
Summary assessment of managing the project dynamics

Instructions
For each of the factors listed below, circle one number in each of the two scales provided to indicate the extent to which the factor is currently met and the effort you are currently making to address the factor. A key for each scale is provided below.

Met?
This scale is to assess the extent to which each factor is currently met:

1 = Not met at all
2 = Met to only a slight degree
3 = Met quite well
4 = Fully met

Effort?
This scale is to assess the effort you are currently making to address each factor:

1 = No effort at all
2 = A little effort
3 = Quite a lot of effort
4 = Considerable effort

Factors	Met?	Effort?
D1. Teams should establish project constraints at an early stage	1 2 3 4	1 2 3 4
D2. Teams should establish programme, highlighting critical dates	**1 2 3 4**	**1 2 3 4**
D3. Teams should agree procedures and methods of working	1 2 3 4	1 2 3 4
D4. Teams should allow adequate time to assess client's needs	**1 2 3 4**	**1 2 3 4**
D5. Teams should validate information with the client organisation	1 2 3 4	1 2 3 4
D6. Teams should improve feedback to all parties throughout project	**1 2 3 4**	**1 2 3 4**

Instructions (continued)
When you have done this for each of the factors, use the two scores to locate each factor, using its list number, in the grid provided to the right of this box.

Entries that fall in the black areas can be ignored. For the remainder, the lighter the area the more urgent it is for there to be a concerted effort on the factor concerned. In this way high priority factors can be identified.

These high priority factors can be combined with similar results for the other 'five-box' areas in Table 12.1 on page 133.

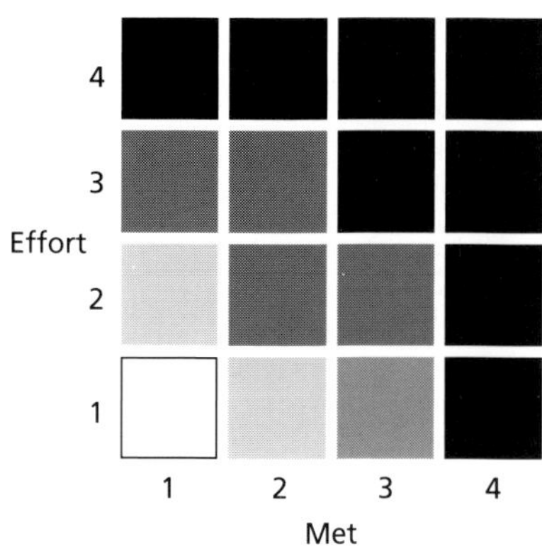

Summary assessment grid

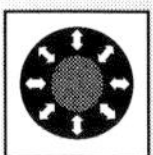

4
Appropriate user involvement

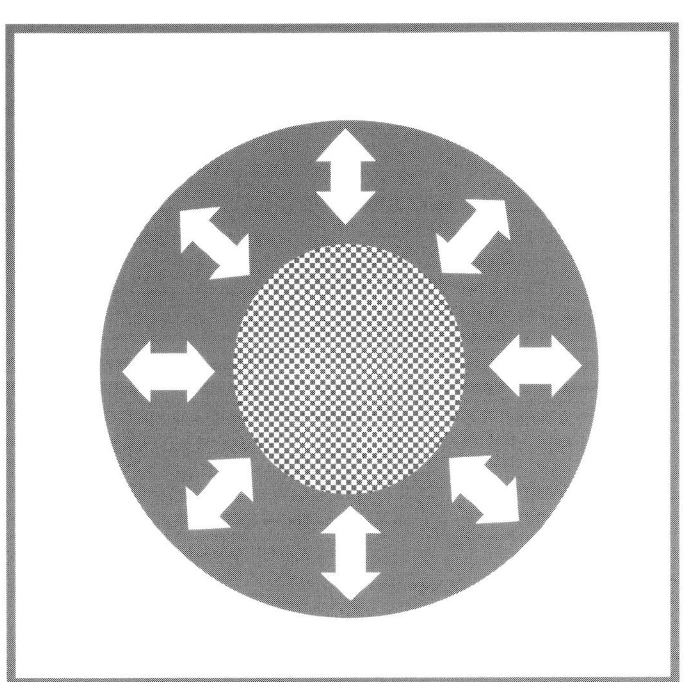

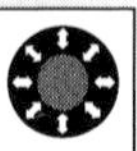

4
Appropriate user involvement

Discussion

Sometimes the client is simple to conceptualise. Let's say it is a private individual who wants to do work to their own property. But, even then they will have a spouse and maybe children, whose views will probably be divergent. With larger client organisations the problems are compounded and it is essential to make a distinction between the perspectives of 'paying clients' and 'end users'. This is illustrated in Figure 4.1 (Zeisel, 1984). The diagram indicates that a relatively clear relationship usually exists between designers and paying clients, but that this can leave important gaps in understanding between the clients and the end users and between the designers and end users. It could be argued that this does not matter and in a limited sense a designer may satisfy a paying client without taking these other factors into account, but the satisfaction is likely to be short-lived. Without engaging end users in the process, an important stimulus to the creative process of design is lost and the client is likely to be left with long-term dissatisfaction amongst the very

Zeisel, J. (1984) *Inquiry by Design*, Cambridge University Press, Cambridge.

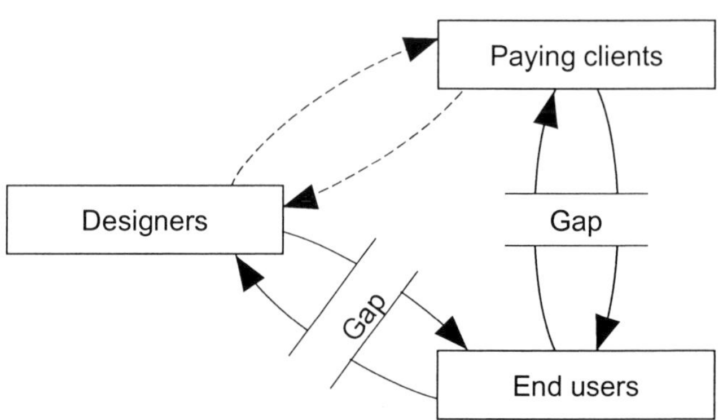

Figure 4.1
Closing the gaps. Source: Zeisel, 1984

people they would wish to support through the built solution being developed.

The implications of this tension can be clarified by considering the implications for the style of consulting that is employed. Three alternatives are given in Figure 4.2 (Garrett, 1981). Expertise consulting stresses the relationship between the consultant and the client excluding the problem owner who would typically be the end user. In this scenario the consultant is employed to provide solutions that satisfy the client. This is in contrast with process consultancy where a consultant interacts extensively with problem owners in order to clarify their needs. This can give rise to very different possibilities from expertise consulting and the consultant is acting more in listening mode rather than imposing solutions from outside. The disadvantage of process consulting is that the synergies from taking an organisation-wide view can be lost and the client may feel excluded and disenfranchised which is dangerous for a consultant if the client ultimately holds the purse strings.

Garrett, R. (1981) Facing up to change, *Architects Journal*, 28 Oct., pp. 838–42.

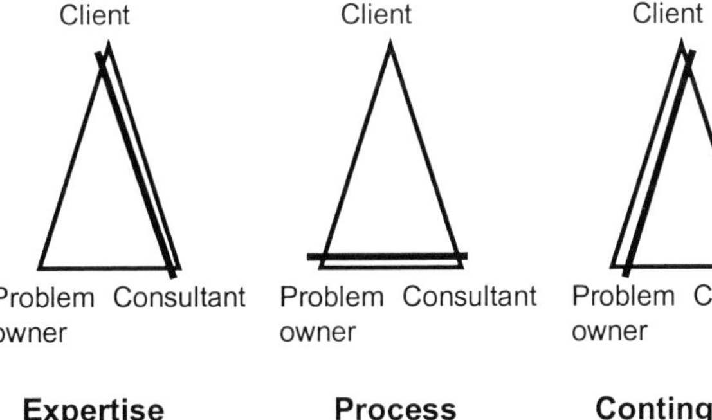

Figure 4.2
Styles of consulting. Source: Garrett, 1981

Contingency consulting is another variant where a consultant arbitrates between the client and problem owner on the assumption that most of the information needed for an optimum solution already exists within the organisation. Garrett presents these as alternatives and tends to favour contingency consulting. An interesting insight to this view was provided through some work with the property arm of a large bank. Their staff made the point that they in fact start with expertise consulting at the regional level, then engage in

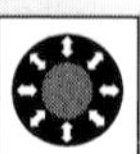

process consulting at the branch level and finally move into contingency consulting in order to resolve differences and reveal synergies to find the optimum solution.

The need for effective interaction between various stakeholders should be apparent

The need for effective interaction between various stakeholders should be apparent, but the difficulty is that clients vary considerably and it is probably better to think in terms of the client system. Aalborg University in Denmark is a good example of this approach. The University was built in five stages and at each stage the design team increased the number of stakeholders that were included in the briefing process. In the final stage the principle was: '*To include everybody who ultimately should be satisfied with the building*'. This led, for instance, to consultation with cleaners specifically about the junction between floors and walls. If consultation of all relevant stakeholders is the objective, the question becomes how to do this effectively? Focus groups have become a popular tool and have great potential, but can easily cause as many problems as they solve.

Focus groups . . . can easily cause as many problems as they solve

One major problem is managing the motivational swings of such groups. Simply in asking the question '*What would you like to see provided?*' expectations are raised which can probably never be satisfied. As a consequence (see Figure 4.3; Barrett, 1995) the motivation of the focus group falls rapidly as they realise the effort and complexity of the problems faced and a very negative atmosphere can prevail, which only clears when the 'light at the end of the tunnel' becomes visible. It may be that at the end of the project they do feel

Barrett, P.S. (ed) (1995) *Facilities Management: Towards Best Practice*, Blackwell Science, Oxford.

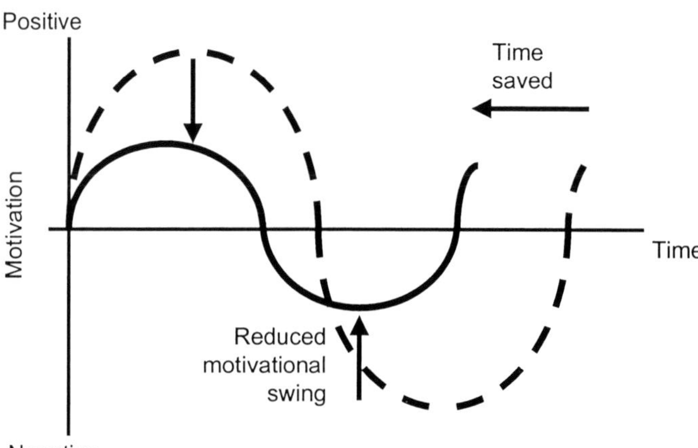

Figure 4.3
Managing uncertainty. Source: Barrett, 1995

that what they have is better than before, but in the meantime a lot of conflict has usually taken place and a lot of effort has been dissipated in trying to cope with the organisational politics underpinning this conflict. There is evidence that if expectations can be dampened down to some degree at the start then the almost inevitable trough can also be minimised and the overall process reduced, probably in calendar time terms but almost certainly in terms of the resources and effort that has to be expended overcoming self-inflicted problems from the process itself. The solution does not have to be complicated. If the initial question is '*What would you ideally like to have? We probably won't be able to provide it, but of course we will do our best*' then the whole process is set in train on a much more realistic, but still positive, basis.

Another key problem area with focus groups is the impact of group dynamics. An example was given in Chapter 1 of the 'text book' use of focus groups which had very negative impacts because various people felt excluded. This can happen within focus groups as well as between those in focus groups and the rest. A very famous experiment was done in the 1950s (Asch, 1955) which illustrates the power of social pressures. Individuals were shown two cards with different lengths of lines drawn on them as shown in Figure 4.4. They

Asch, S. (1955) Opinions and social pressures, *Scientific American*, November, 31–35.

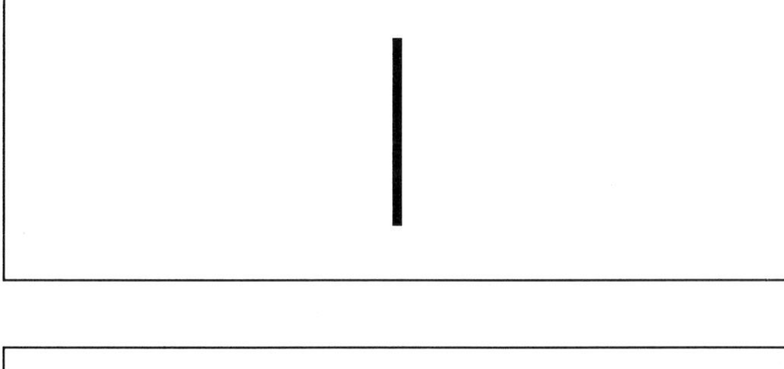

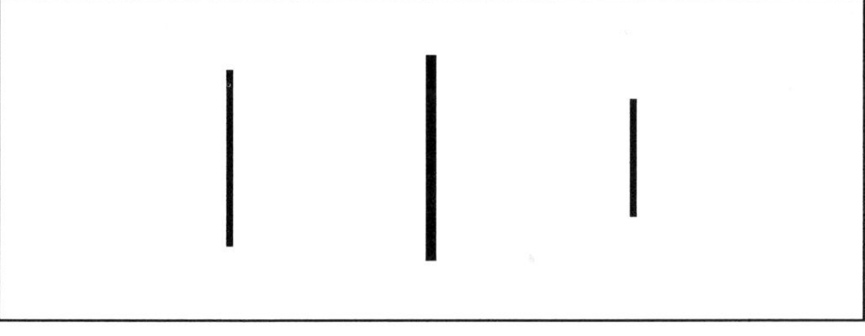

Figure 4.4
Asch's experiment. Source: Asch, 1955

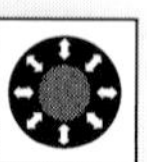

were asked which line on the second card was the same length as the line on the first card. This was not a trick question – it was obviously the middle line. Anyone asked this question on their own got it right. The situation was then rigged so that two people were asked the question and one of them purposely gave the wrong answer. The guinea pig was not swayed by this and consistently continued to give the right answer. The experiment was repeated with groups of three in which two people were giving the wrong answer. In these cases 14% of the guinea pigs changed their view to agree with the others and those who didn't were really slightly jumpy! The experiment was repeated with groups of four with three other people giving the wrong answer. In this case, on average, 32% of the guinea pigs changed their view and those that didn't were extremely uncomfortable. The fact that those who adapted to meet the group view were probably simply doing it to avoid embarrassment is of little comfort to those who have to manage in practice where behaviour is the thing that ultimately counts.

there is a real danger that dominant voices . . . will create an environment in which other people find it difficult to express contrary views.

In the briefing context there is a real danger that dominant voices or strong groups within focus groups will create an environment in which other people find it difficult to express contrary views. It is quite common for this to show up in the corridors after a meeting when people say 'I didn't really agree with that but didn't like to say'. But by then it is too late. The result is that a broad range of information and views is not properly elicited and a lot of bad feeling and disenfranchisement is generated.

the right people can be asked the right questions in the right context

Appropriate user involvement would appear to argue for using issue-specific groups in a flexible way. Thus the right people can be asked the right questions in the right context. For instance, big differences in seniority within a group would be avoided and it would become possible to ensure that the debate involved is of relevance and more than passing interest to most of the people, most of the time. This could be taken a step further, if justified, by interviewing individuals within their working situation. Whichever approach is taken, meaningful feedback as to actions taken must be provided to maintain the goodwill necessary for effective communications.

Case studies

With regard to 'appropriate user involvement', the following main factors should be considered by clients in conjunction with construction professionals.

- *User involvement benefits should be understood*
- *User involvement should be assessed relative to each situation*
- *User participation should be planned to allow relevant data to be collected*
- *User group dynamics should be considered*
- *User involvement should be maintained throughout the project.*

User involvement benefits should be understood

Involving users in the briefing process can potentially have a number of different benefits. Users are the experts in their areas and so it makes sense to consult them. Managers will probably have an overall appreciation of what occurs and should be consulted about broad concerns. Only users appreciate the fine details of their work and so they are best placed to ask for changes that could make real improvements to the way they work.

There may be other indirect benefits. If users are asked for their opinions they will feel part of the decision-making process and will 'own' the project, so even if they do not get everything that they asked for they are still likely to be more satisfied with the final outcome. Associated with this is the feeling that management believes that the opinions of the staff matter and that they have something to contribute to the running of the organisation.

User involvement should be assessed relative to each situation

Organisations vary in their attitudes as to whether users should be involved in the construction process. Some try to involve all the affected users, others only involve managers. In order to assess the most appropriate level of user involvement in each case, some consideration should be given to the culture of the client organisation. For example, do senior management place importance on the opinions of their staff

'The philosophy of the client is very, very much to involve everyone, and they really mean everyone, staff, visitors, down to the local people. We have 32 user groups which is potentially a nightmare.'
(Project manager)

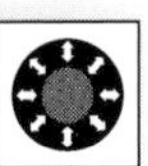

and work with them? Or do managers make all of the decisions? If managers are not likely to support and follow-up on users' requirements then users should not be asked for their preferences. Raising users' expectations falsely is likely to do more damage than not consulting them at all. If the final result does not reflect their requirements they are going to resent wasting their time.

'But what you would not do is make sure every person in every department had their say as this would be a totally unmanageable exercise timewise and you cannot please all of the people all of the time. You have to get the managers to take a rational view.'
(Client's project manager)

The size of an organisation is also likely to have a bearing on the amount of user participation. Some organisations try to involve all users, but this is only practical when an organisation is small, otherwise the briefing stage will take a long time and again people are likely to be disappointed if their requirements are not included.

In some instances the occupants of a building may change on a regular basis. In this situation there is probably no gain in asking individual users about their specific requirements, as the next occupants may have totally different needs. However, it may still be worthwhile consulting users about general issues.

'You can always talk to the equivalent people in another hospital. A matron in one hospital has pretty much the same problems and knowledge as another matron. If you cannot do something similar in your own company, you can do it with another. People should always have contacts within their own field.'
(Client's project manager)

In some of the case study organisations it was not always possible to consult with the users of the building. In some instances new buildings were needed because of organisation expansion and would be occupied by totally new staff. In other instances managers believed that they knew everything about their departments and refused to consult with their staff. If it is not possible to consult with existing or known users, the design team may find it useful to consult people with similar jobs.

Good practice

In one of the case studies a new hospital was being constructed for a private hospital group and staff would not be appointed until the building was complete. In order to compensate for the lack of users, the designers consulted with staff in a comparable hospital to get a feel for what would probably be needed.

User involvement should be planned to allow relevant data to be collected

Users who become involved with construction projects will also normally carry on with their day-to-day routines and so their workloads will increase. It is important that their organisation and the design team realise this and allow them enough time to consider the construction project in as much detail as is required. Users need to feel that the project is important to the organisation and their managers, otherwise they will not dedicate enough time to it.

In one of the case study examples, the users were invited to attend progress meetings so that they could remain informed about the scheme and raise any concerns. This was potentially a good idea, but these meetings were held at 5:30 PM and so attendance was poor. The users felt that the timing of the meetings gave an indication that senior management did not really want them to comment on the scheme as they were not allowed time off from their normal duties.

'The user groups are given a two page view on what they are there for and what their remit is. We also give them a page of A4 as their reporting mechanism, with 32 user groups we obviously don't want a large document from each group. They can talk a lot, but I don't want to hear about all of it. I just want to know what they really want to change.'
(Architect)

Few organisations build on a regular basis and so it is quite likely that users will be unfamiliar with the construction process. It is important that such users are briefed properly on what is expected of them and what will happen as the project progresses. They need to be aware that the final design will be a synthesis of many different requirements and so not every preference will be included.

'We started off by getting the sixty users in the large auditorium. We said this is the design process we are going to go through. We thought by taking them through it, warming them to what we were trying to do, they would understand the process and the timescales. Very often people say, "What is the difficulty, knock a few walls down, do it over the weekend!' It's only when you show people the processes that they realise why it takes so long.'
(Architect)

Good practice

Before they meet with the design team users should have preliminary meetings to discuss their requirements and agree their priorities. Otherwise the consultants will find themselves listening to internal disputes which will not necessarily have anything to do with the final design.

All involved users should be briefed on their responsibilities and what will happen as the project progresses. It is a good idea to brief users collectively if possible, then they will all be working on the same basis. This briefing could take the form of a presentation or a standard document.

Different users should be involved at different stages of the briefing process, so that the most appropriate people are

'The first time we had a meeting with a user group it was a farce, because they were thinking and talking on their feet which just didn't work. So I said, "This is not on, this is the information we want you to go away

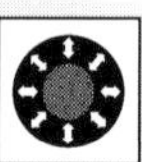

and talk about. When you have made your minds up come and talk to me about it." The next time was much better.'
(Client's project manager)

'The reality is that people learn things from user groups rather than actually contribute to them. They will cause mischief after the meeting by lobbying for a particular cause. I've been at meetings where I ask if anyone has any problems and I just watch the noddings dogs. You know very well that there will still be problems, people get together in small groups afterwards and say it doesn't work like that! So we keep the groups small, because people will talk and are prepared to put ideas forward.'
(Project manager)

'I was showing a group of users around their new building. People would ask questions and I would say that I had told their representative what was going to happen, but they hadn't heard anything. So they had been subjected to a lot of unnecessary rumours and fears.'
(Project manager)

consulted on each issue. For example, during the initial stages managers will be able to provide guidance on broad issues. Other staff should be consulted later on as they will be more aware of the detailed requirements.

User group dynamics should be considered

When considering the mix of people in a user group it is important to obtain a balance of views, so that the final design will not be biased towards one sub-set. If managers, or strong personalities, are included in groups they may exert unseen influence over the other members who do not want to contradict them in public. This occurred in one of the case studies where a senior person within the department attended a user group meeting and dominated the conversation. His opinions remained unchallenged, then he left and everyone else said that they didn't agree with what he had said. This left the architect in a very difficult position as he did not know who he should be listening to.

Users need to have confidence that their representatives will pass on their requirements as necessary to the design team. The case studies, however, highlighted many occasions when the user representatives failed to perform their duties correctly. The representatives did not always pass on all the relevant information as they often unknowingly filtered information that they thought was irrelevant or unimportant. Worse still some representatives deliberately withheld information as they believed this gave them some sort of power over the other users. However, the most common problem was that representatives did not report back to the other users once decisions had been made.

Good practice

User group representatives should ideally be selected by the people they are representing, i.e. their departments or similar, rather than by managers. This should give the other users confidence that their views are being passed on.

User representatives should be briefed properly by the design team and client so that they are fully aware of what they are expected to do and what the reporting mechanisms are.

It may be a good idea to formalise the procedures so that representatives take written documents approved by the other users to meetings with the design team, rather than just reporting verbally on users' requirements. This will also help the design team, as users will have to discuss their requirements in more detail in order to write them down.

Once representatives have attended user group meetings they need to formally report back to the other users so that they are kept informed of progress. This could be done either in a written document or at a short briefing meeting.

User involvement should be maintained throughout the project

Users often complain that once they have passed their requirements on to the designers they hear very little about the project, so they are unsure of how the project is progressing and what the final building will be like.

Feedback should be given to users after they have provided the design team with information, so that they can verify that their requirements have been interpreted correctly and also get an idea of what can realistically be incorporated into the design.

Users should be shown around a project before completion, so that they can see what is happening and where they will be located. If users become familiar with the layout of a building it will be less of a shock when they move in. They can also begin to plan how to make best use of the space and so operate more effectively as soon as they move in.

'There was a period early on when groups of staff were taken around the site. Now it is more enclosed it would be a good idea for them to be allowed around again, as they would get a better impression. That is something that would do a great deal for staff morale, actually to be able to get in there and have a look at it.'
(Senior user)

'They came up in luxury coaches, they were shown around by me and I answered every single question. At the end before they had their big lunch, we showed them a workstation all set up. They said, "Yes, but this is for the managers", I replied, "No everyone is getting one". So they went away on a bit of a peak. The whole thing was strategically and psychologically worked out. They didn't want to relocate, so we had to convince them that everything would be OK.'
(Client's project manager)

Good practice

In one project the design team elicited feedback by inviting all the users to a meeting where they were shown enlarged concept drawings. Users were free to comment on the proposed layout and the architect wrote everyone's remarks directly on the drawing. This allowed the design team to obtain an enormous amount of feedback very quickly. It also meant that each department could see how their

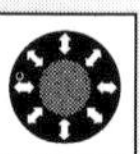

requirements would interact with other parts of the scheme.

Summary and self-assessment

Overall this chapter has stressed the importance of consulting various stakeholders and balancing their wishes and desires with what is physically and financially possible. It is a question of asking the right people the right questions, at the right time.

Table 4.1 focuses on the main factors necessary to achieve this end and provides a self-assessment tool for the reader to assess where effort is most urgently needed. The tool can be used for the reader's organisation in general, some part of it, or it could be used to assess a specific project and just those people involved.

The outcome of the analysis is the identification of those factors that are a high priority. An explanation of how the table is completed, and how the resulting information can be used, is given at the end of Chapter 2.

It should be remembered that the factors highlighted at this stage are in only one area. There are four others, which together make up the 'five-box' model. It may be worth waiting until similar analyses have been done for the other areas and a combined summary created, as shown in Chapter 12. It may then be even easier to link together high priority factors with issues that are important and urgent for the firm.

Table 4.1
Summary assessment of appropriate user involvement

Instructions
For each of the factors listed below, circle one number in each of the two scales provided to indicate the extent to which the factor is currently met and the effort you are currently making to address the factor. A key for each scale is provided below.

Met?
This scale is to assess the extent to which each factor is currently met:

1 = Not met at all
2 = Met to only a slight degree
3 = Met quite well
4 = Fully met

Effort?
This scale is to assess the effort you are currently making to address each factor:

1 = No effort at all
2 = A little effort
3 = Quite a lot of effort
4 = Considerable effort

Factors	Met?	Effort?
U1. User involvement benefits should be understood	1 2 3 4	1 2 3 4
U2. User involvement should be assessed relative to each situation	1 2 3 4	1 2 3 4
U3. User participation should be planned to allow relevant data to be collected	1 2 3 4	1 2 3 4
U4. User group dynamics should be considered	1 2 3 4	1 2 3 4
U5. User involvement should be maintained throughout project	1 2 3 4	1 2 3 4

Instructions (continued)
When you have done this for each of the factors, use the two scores to locate each factor, using its list number, in the grid provided to the right of this box.

Entries that fall in the black areas can be ignored. For the remainder, the lighter the area the more urgent it is for there to be a concerted effort on the factor concerned. In this way high priority factors can be identified.

These high priority factors can be combined with similar results for the other 'five-box' areas in Table 12.1 on page 133.

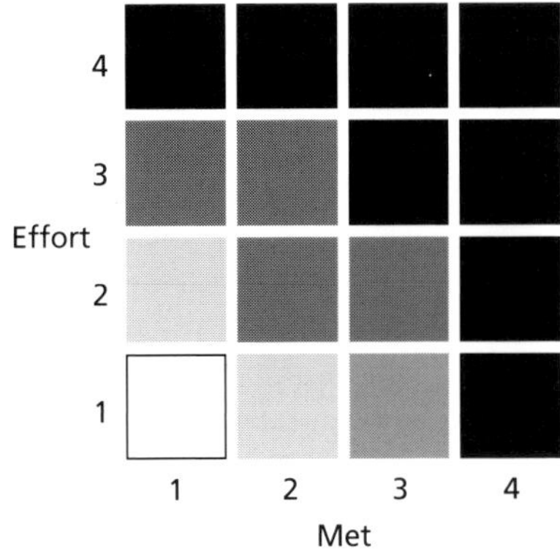

Met

Summary assessment grid

5
Appropriate team building

5
Appropriate team building

Discussion

Different projects need different sorts of teams and as we have already seen different clients need different sorts of support. Focusing on professional firms, different types of firm can be seen to be determined by the relative proportions of 'finders', 'minders' and 'grinders' as shown in Figure 5.1 (Maister, 1982). For a well-determined construction project, where much of the work can be proceduralised, based on past experience, a 'procedure' firm can be the ideal and most economic choice. At the other extreme, where an innovative, one-off solution is required to a rather non-standard situation, a 'brains' firm may be needed with its high proportion of senior staff. For something in between a 'grey-hairs' firm may be preferable. So the ability of a firm to either do the specific sort of work in hand or to solve open-ended problems can be a relevant way of choosing team members, depending on the degree of innovation required.

Maister, D.H. (1982) Balancing the professional service firm, *Sloane Management Review*, MIT, Fall, 15–29.

the ability of a firm to either do the specific sort of work in hand or to solve open-ended problems

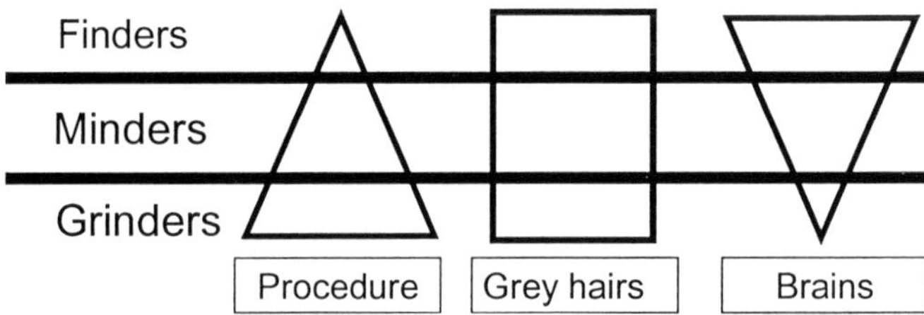

Figure 5.1
Maister's typology of firms. Source: Maister, 1982

As a counter-intuitive example, one large pharmaceutical company that had had many laboratories built, purposely chose an architect for one project with no experience in the pharmaceutical field, but a brains-type structure and a reputation for innovation, in order that, on this one project at least, new ideas would re-energise their extensive collected, but self-reinforcing, experience.

Although companies are one building block, in practice success tends to depend on specific individuals gelling. One way of categorising individuals springs from their emphasis on different parts of the learning cycle (Kolb, 1976, building on Lewin, 1951) shown in Figure 5.2. These tendencies are individual, but may tend to follow employment group, as shown in Figure 5.3 where they have been applied to construction (Powell, 1991). From this it can be seen that architects tend to be sensor-doers, lighting/sound engineers watcher-thinkers and quantity surveyors watcher-thinker-doers. Strangely, as James Powell points out, marketing managers seem to 'watch' and then 'do', without 'thinking' in between! The important thing is that the whole learning cycle is represented in the team, given that it seems very unusual for any one individual to have a full spread. Thus, in parallel with ensuring the necessary technical knowledge is accumulated within the team, complementary learning emphases are also desirable.

Lewin, K. (1951) *Field Theory in Social Sciences*. Harper and Row, New York.

Kolb, D.A. (1976) *The Learning Style Inventory Technical Manual*, McBer, Boston.

Powell, J.A. (1991) Clients, designers and contractors: the harmony of able design teams. In *Practice Management: New Perspectives for the Construction Professional*, (eds P.S. Barrett and A.R. Males), pp. 137–48, E&FN Spon, London.

This second aspect is quite hard to assess in practice and one of the clearest indicators is effective past performance in various relationships. This is because much of the effectiveness will be built up in informal processes, shared assumptions and positive sentiments. There is an argument

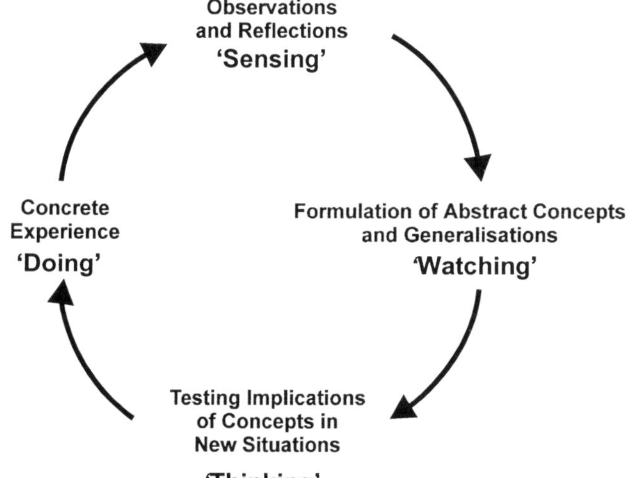

Figure 5.2
Lewin's/Kolb's learning cycle

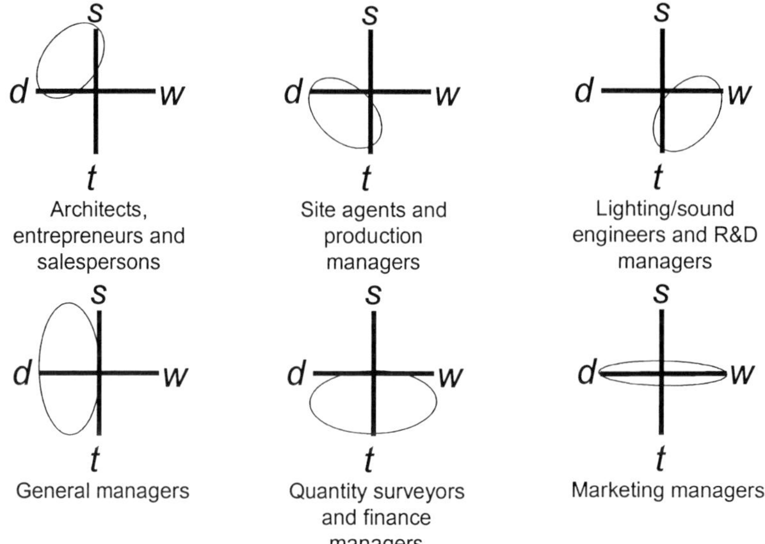

Figure 5.3
Learning styles and construction. Source: Powell, 1991

here for keeping together core teams that have worked effectively in the past, possibly with some new element to keep the cocktail bubbling. The benefits of this sort of team's stability in the turbulent construction environment are evident when the steps in forming an effective group are considered, namely: forming, storming, norming and, only then, performing (Handy, 1985). In construction it is common experience that teams are thrown together – they 'form'. There is then a period of 'storming' as individuals get to know each other and their aspirations are, hopefully, accommodated into the overall project objectives. This is then reflected in a period of 'norming' where agreed ways of working and joint goals are created. This achieved, the group can properly 'perform'! But how often do construction teams get the chance to go right through the process before the project is finished? Many seem to stick at 'storming'. Alternatively, the storming and norming phases are missed out so there is no real foundation for effective performance. Clearly by keeping some key members of the project team together it is possible to perform at a high level, rapidly. This is not a panacea for appropriate team building,

Handy, C.B. (1985) *Understanding Organisations*, Penguin, Harmondsworth.

Many [construction teams] seem to stick at 'storming'

but one dimension that can be seriously considered along with the general types of firms and individuals needed to provide the client organisation with what lacks, so that the client's aspirations can be achieved.

Case studies

When considering 'appropriate team building' clients and construction professionals should focus on the following points.

- *Team selection should focus on skills, not just financial considerations*
- *Teams should approve and understand the management structure of projects*
- *Team members should remain consistent throughout projects.*

Team selection should focus on skills, not just financial considerations

When clients select the project team they need to take a number of factors into consideration. Cost should not necessarily be the driving factor. Clients need to also think about whether they will actually be able to work with the consultants and how compatible the different parties are.

Clients should take the time to find out about consultants' previous work. For example, if a client wants a building constructed from traditional materials, they should ensure that they do not appoint an architect whose buildings are all glass and metal. Clients should also try to ascertain what the balance of power will be. If the client is inexperienced will the consultants make all of the decisions or will the client be expected to take the lead?

During the case studies clients sometimes complained that they felt that the consultants were using a standard design solution, rather than one tailored to their own particular organisation. Hence clients should try to ascertain that the consultants will listen to their requirements and that they have no preconceived ideas. In order to ensure that they were getting a tailor-made solution, one of the clients

'So we selected each of the consultants individually on this project. The problem with that is you have to make sure they can all work together. It is just like interviewing a new member of staff, the important thing isn't just what they have done, but will they fit in with everyone else. Projects are very much focused on people getting on and how communication works.'
(Project manager)

'Typically we would go through: practice experience, specific hospital experience within the past five years, projects of a similar size and complexity . . .'
(Project manager)

specifically appointed consultants with no previous experience related to the organisation's particular field. These consultants were then forced to study the organisation in minute detail so that they could design an appropriate building.

'All the laboratory specialists that we interviewed, about four, you knew exactly what they were going to do and that they had a lab designed in their head already.'
(Client's project manager)

When selecting a consultant organisation, clients need to check which specific members of the consultant company will actually work on the project. It is quite possible that a lot of the work will be carried out by less senior members of staff. If this is the case how will their work be supervised?

'So we looked at it from the practice point of view, but more importantly we looked at it with regard to the proposed team of individuals within each firm. Who were they going to use and what had they done before? We looked at the management structure of the internal team, their approach, how they were going to communicate, speed of response, how they would react to client change and instructions.'
(Project manager)

Good practice

Experienced clients often tend to use the same consultants, or group of consultants, for similar buildings, as they feel that this speeds up the design process. The consultants are already familiar with the working practices of the organisation and can begin designing quite quickly once they have established that the client wants a similar solution as previously. These clients ensure that they will be working with the same people from the consultants, as it is the individuals who have built up the knowledge rather than the organisations.

Teams should approve and understand the management structure of projects

'Communication is never perfect, I have never been on a job where it was, we are as much to blame as anyone else. There have been a few instances where communications between the design team have not worked very well and the project manager has made his views quite clear that he expects it to be better than that. I think it is generally just a lack of awareness of the programme and what the weekly deadlines are.'
(Quantity surveyor)

Another factor to be taken into consideration is how will the different consultants work together. Team members should be sure that they understand fully the management structure of the project. In several of the case study projects it was not always clear who was responsible for what and so some activities were duplicated while others were neglected altogether. Similarly there were occasions when friction occurred as consultants, most often architects and project managers, appeared to disagree over who should be directing the project and who the client should be looking to for advice. Traditionally architects have been the lead consultant and have seen themselves as the client's major advisor. However, with the growth of the project manager, the architect's role as decision maker has been somewhat usurped.

Project members also need to be working towards the same goal and people should be dissuaded from looking after their own interests. In one of the case studies a consultant appeared to be pushing an alternative procurement method. It transpired that another part of their organisation were experts in this method and so it was possible that they would obtain further work if this method was employed.

As part of the project requirements some of the project teams were obliged to produce documents outlining procedures and responsibilities, which should have eliminated any confusion. On one occasion this document was not completed until the detailed design work was underway and so was not really of much use. However, the roles of the different parties had become particularly confusing during this project and so this document should have been taken more seriously.

'I think the designers, both the architects and the engineers, committed themselves to fairly tight design periods. We were very dubious from the start that it was going to be achievable and we made our views known to the project manager. He said that that was what the client laid down in the brief and so that is what we had to work to. They (note – not we!) didn't achieve it for some good reasons and some bad reasons. I think they were just too optimistic that they could achieve it.'
(Quantity surveyor)

Good practice

All of the consultants and the client should be asked to define and record their responsibilities on paper. These descriptions should be circulated for comment and then everyone should meet to discuss, rationalise and finally agree on what should be the responsibilities of each party. These statements should then be formally recorded to avoid confusion.

Team members should remain consistent throughout projects

If possible project team members should remain constant throughout a project. In several of the case studies the representatives from the different consultant organisations altered from meeting to meeting. This meant that people were often not up to speed on how the project was progressing and so they had to be briefed during project meetings, which wasted time. In addition, each new representative tended to have a different view of the situation and alternative approaches were often suggested. The project teams often felt obliged to consider these new ideas and so project progress was affected.

Project teams should also consider which consultants

'We have what I term a peer group meeting once a month. It is attended by the client and principals from each of the design team consultants. By principals I mean the people who are the level above the people who are actually doing the job. If they can't attend, the next person up the tree (i.e. a more senior manager) comes along. These meetings consider in summary the progress of the project and future policy. We also talk about

any personnel problems that may be occurring either with any member of the team or relationships between people. I believe the value of these meetings is that they involve senior people from the consultant organisation. Too often normally you find that they know nothing about the project that their staff are working on.'
(Project manager)

should become involved at which stage of the project. Many projects have a core of consultants who are involved throughout the project and then other consultants are brought in to deal with very specific parts of the building during the detailed design stage. It should be remembered, however, that these consultants may require a specific set of conditions in order to achieve their contribution. For example, a services consultant may need voids in specific locations to allow for cable runs, etc. In some of the case studies specialist consultants (or users) were introduced at what was considered an appropriate time in the project, but their input had an unforeseen impact on the design and this resulted in major reworkings of the schemes. Hence, where possible, all consultants should be involved from an early stage, at least to an extent, so that their requirements can be considered and incorporated.

Summary and self-assessment

Overall this chapter has stressed the importance of getting a complementary mix of skills and attitudes to meet the project demands.

Table 5.1 focuses on the main factors necessary to achieve this end and provides a self-assessment tool for the reader to assess where effort is most urgently needed. The tool can be used for the reader's organisation in general, some part of it, or it could be used to assess a specific project and just those people involved.

The outcome of the analysis is the identification of those factors that are a high priority. An explanation of how the table is completed, and how the resulting information can be used, is given at the end of Chapter 2.

It should be remembered that the factors highlighted at this stage are in only one area. There are four others, which together make up the 'five-box' model. It may be worth waiting until similar analyses have been done for the other areas and a combined summary created as shown in Chapter 12. It may then be even easier to link together high priority factors with issues that are important and urgent for the firm.

Table 5.1
Summary assessment of appropriate team building

Instructions
For each of the factors listed below, circle one number in each of the two scales provided to indicate the extent to which the factor is currently met and the effort you are currently making to address the factor. A key for each scale is provided below.

Met?
This scale is to assess the extent to which each factor is currently met:

1 = Not met at all
2 = Met to only a slight degree
3 = Met quite well
4 = Fully met

Effort?
This scale is to assess the effort you are currently making to address each factor:

1 = No effort at all
2 = A little effort
3 = Quite a lot of effort
4 = Considerable effort

Factors	**Met?**	**Effort?**
T1. Team selection should focus on skills, not just financial considerations	1 2 3 4	1 2 3 4
T2. Teams should approve and understand management structure of projects	1 2 3 4	1 2 3 4
T3. Team members should remain consistent throughout projects	1 2 3 4	1 2 3 4

Instructions (continued)
When you have done this for each of the factors, use the two scores to locate each factor, using its list number, in the grid provided to the right of this box.

Entries that fall in the black areas can be ignored. For the remainder, the lighter the area the more urgent it is for there to be a concerted effort on the factor concerned. In this way high priority factors can be identified.

These high priority factors can be combined with similar results for the other 'five-box' areas in Table 12.1 on page 133.

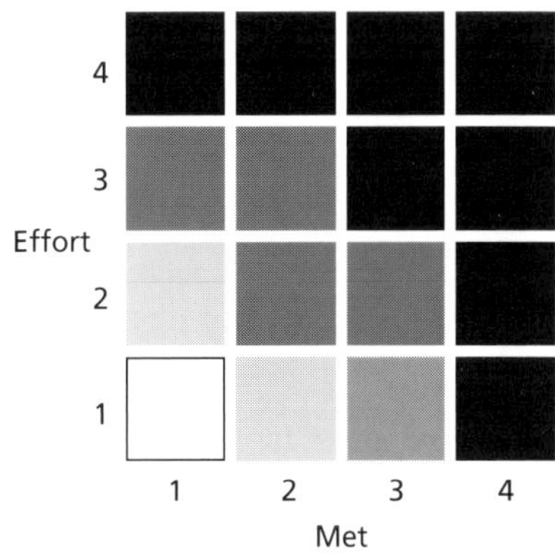

Summary assessment grid

6
Appropriate visualisation techniques

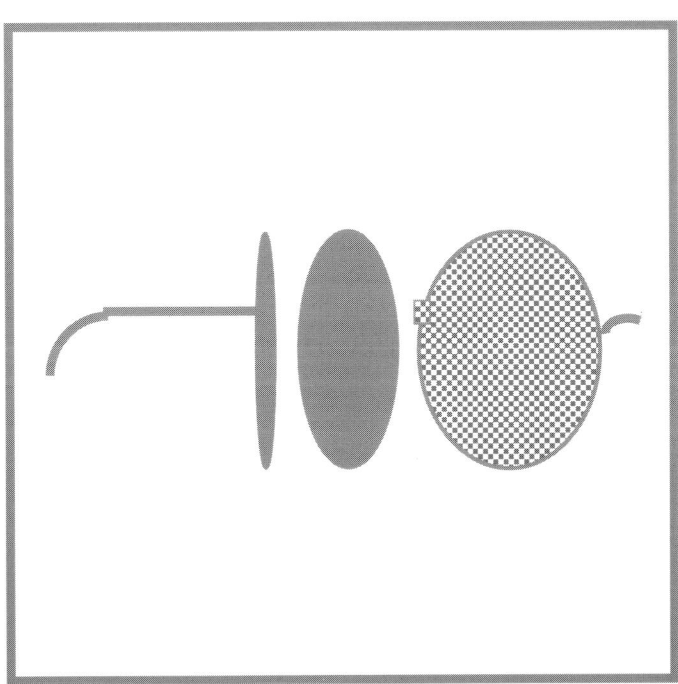

6
Appropriate visualisation techniques

Discussion

Effective communications are needed between all parties in order to achieve the optimal result.

The briefing process is essentially one of communication. Effective communications are needed between all parties in order to achieve the optimal result. However, in this context the problems revolve around a specific aspect of communication which arises in construction because of the very tangible nature of the final product. When the project is finished a client will have a built artefact which they can touch, smell, walk around and see from many different angles. In contrast at the start of the process they will only have a broad idea of what is wanted and what they are likely to get. The design team will provide information, typically in terms of drawings. The problem is that very few clients can really read drawings in a meaningful way.

'Help in escaping this flat land is the essential task of envisioning information – for all the interesting worlds that we seek to understand are inevitably and happily multi-variant in nature. Not flat lands.'
(Tufte, 1990)

Tufte, E.R. (1990) *Envisioning Information*, Graphics Press, Connecticut.

Explanation of basic communications concepts can be found in any standard management text, for example:

Megginsson, L.C., Mosley, D.C. and Pietri, P.H. (1989) *Management Concepts and Applications*, 3rd edition, Harper and Row, Cambridge, Mass.

Communication theory can immediately throw some light on this problem. It is generally held that for effective communication four strategies should be pursued. First, the information should be *coded* to make sense to the recipient. Second, the message should be *robust* and then *reinforced*, ideally by repetition and the use of multiple media. Fourth, *feedback* should be sought to ensure that the receiver has received what the sender thought they were sending. There is a fifth element, which is that the whole *context* of the communication process needs to be taken into account.

the message is often coded, but not always in the language that the recipient will understand.

When the communication techniques used in construction are considered, it is clear that the message is often coded, but not always in the language that the recipient will understand. The very terminology of construction leaves many construction clients in a wilderness and little effort is made to put the proposals in familiar terms. For instance, on something as simple as sizes of rooms, it is very difficult for many clients to appreciate the information contained on drawings. However, simply walking the client into a room of the same size, or taping it out on the floor would immediately get the information over. The same approach can be used by taping proposed glazing bars onto windows, chalking windows onto existing walls, and so on. This idea of doing more than the minimum (relying on drawings) links to the communications concept of providing sufficient 'redundancy' so that if only some of the information is communicated the message still is sufficiently robust to come over. A trivial, but telling, example of this *not* happening between construction professionals was provided by a project where specialist mechanical and electrical services were shown using different coloured lines, and then the information disseminated by black and white photocopies! If the lines had also been drawn differently using dots and dashes then the information would not have been so fragile.

'Reinforcing' the message reflects an aspect of the section on managing project dynamics above. If undue reliance is placed on the clients' initial, probably faulty, perception of what is meant, problems will ensue. If they are given further clarification and so come to see the proposals more clearly this is very likely to be beneficial. The interactive aspect of this is reflected in the feedback element of effective communications. An overriding problem with many attempts to communicate in briefing is exemplified by the comment reported in Chapter 1, that '*clients don't understand construction*'. In fact construction is a world with its own terminology and processes, many of which are taken for granted. Without a significant effort it is unreasonable to expect many clients to really understand what is being proposed.

Case studies

When project teams are considering 'appropriate visualisation techniques' they should focus on the following factors.

- *Visualisation techniques should be employed to increase potential for shared understanding*
- *Visualisation techniques should be adequately resourced*
- *Visualisation techniques should be used effectively.*

Visualisation techniques should be employed to increase potential for shared understanding

'They couldn't read drawings of course, that was another thing! They had no idea of sizes on drawings until you actually taped it out on the floor. We did that with everybody to make sure they were happy.'
(Project manager)

The aim of a construction project is to produce a building that fits the client's requirements. Many clients complain, however, that the final building is not what they were expecting. One of the major reasons why this occurs is that clients and users, particularly inexperienced ones, find it difficult to translate drawings into an accurate impression of what the final building will be like. Architects, etc. are trained to produce and read drawings, clients are not. A problem that was raised time and time again by the consultants during the projects was the difficulty that clients and users had in reading drawings.

'One thing they could have done easily, and have actually now done a couple of times, is produce images of what the building will be like. It would have helped the staff enormously to get a feel for what it is going to be like inside, rather than just looking at floor plans and elevations which are difficult to read.'
(User, senior management)

In some cases attempts were made to try to assist client understanding of the drawings by using alternative visualisation methods. However, help was not always provided; many consultants just seemed to accept that the problem existed and did not try to overcome it.

Good practice

Consultants should try to assist clients to gain a real appreciation of the proposals by using additional, appropriate visualisation techniques, as detailed below, in conjunction with drawings. The introduction of other media encourages further discussion and may help each participant to gain an increased understanding of what others are trying to achieve. Additionally, by increasing opportunities for discussion, clients are more likely to feel that they are making a real contribution to the project which should increase the level of client satisfaction.

Visualisation techniques should be adequately resourced

Some consultants and clients suggested that it was too expensive and time consuming to use additional methods to aid client and user comprehension of schemes. However, if people are not able to understand drawings there is an increased likelihood of changes being asked for during the construction phase, as they can finally see what they have agreed to. When discussing alternative media many people would think of expensive computer assisted visualisation techniques, such as virtual reality or three-dimensional walk throughs. However, alternative visualisation methods do not have to be expensive as there are some very basic, quick methods which can be utilised to great effect, as listed below.

'Another problem with these architects is that their drawings are illegible. Normally you should be able to read architectural drawings pretty straightforwardly, but even our property department found them difficult to read because it wasn't always clear when lines were walls or other things. It was partly a function of the computer CAD system that they use. Their system works in colour and all the drawings that were produced were in black and white. If you see the colour on the screen or you have got a colour print-out, then it is immediately obvious what the different lines mean, but in black and white it's not. We suggested that we had colour plans, but we have never received any, too expensive I guess.'
(User, senior management)

Good practice

Photographs can be used by the design team during the initial stages of the briefing process to quickly elicit what sort of buildings, materials, spaces, etc. the client likes. Architects could use photographs of their previous work or they could use pictures from journals. It may be a good idea to build up a portfolio of images for this purpose. The design team should not see this technique as a restriction on their creative design skills, but a useful tool that will give them an idea of what sort of client they are dealing with and what sort of design may be acceptable.

Existing buildings can be visited so that clients and design teams can inspect best practice examples and compare alternative solutions. Clients will probably know of other similar facilities that they would like to look at. It is easier to discuss different possibilities with real examples to refer to. Design time may also be saved as architects can adapt workable solutions from elsewhere, rather than re-inventing the wheel. Project teams may find it useful to look at other types of organisations for alternative solutions. Private hospitals, for example, could study hotels, as the design of bedrooms is important in each case.

'The client has paid for relevant hospital staff to see similar specialist hospitals in the States. A lot of the user group co-ordinators liked one hospital in particular which they are constantly referring to. The architect and I are going over to see it for ourselves. They also brought back videos, slides, photos and copies of the architect's drawings. So we have drawn a lot of information from other hospitals and we intend to continue to do so.'
(Project manager)

Taping out room sizes on the floor is a very cheap, quick and effective method of demonstrating real dimensions to clients and users who may not be able to read drawings. The

'With some sticky little issues we actually emptied the atrium space and taped the areas out on the floor to give them an indication of how big things were. Even this wasn't always clear enough, so we put some benches into the marked areas and they could then picture the space. People were bringing bits of equipment that they carried from bench to bench, "No that's too far, that's three steps, I only want two." Then we'd amend the drawings until they were right.'
(Project manager)

location of furniture and equipment should be marked up, or existing pieces brought in, so that clients can test the proposed solutions to see if their requirements are met.

Full size mock-ups should also be considered. A properly constructed mock-up of a workstation may seem expensive, but it is better to get feedback at this stage, rather than when a whole set of inappropriate workstations have been installed.

Design teams should try to use models more as visualisation tools, rather than presentation tools. Site models showing the building in its location are frequently commissioned in the early stages to gain planning permission or to attract funding. Once these goals have been achieved the models are put in glass cases and never touched again. Simple cardboard models are much cheaper to make and are more flexible, as they can be dismantled and rebuilt easily, encouraging real interaction between the designers and the users.

Visualisation techniques should be used effectively

'We are also taking the design team around some of the best UK examples for specific things. Initially we are looking at the use of space, how much space is necessary for certain functions. We can't use the existing hospital because it is amazing what they can do with no space at all. I think the visits are money extremely well spent, although there would be very few people prepared to spend it. One of the greatest benefits from visiting other places is the fact that the clients and design team get to know each other. So later if there is a problem, rather than sending a twenty page letter they make a phone call. From my point of view as a project manager that will save an awful lot of money, but you will never be able to measure it.'
(Project manager)

Some of the projects did employ alternative visualisation techniques, but they were not always used effectively. In one project the architect wanted to demonstrate how the proposed workstations would be arranged in clusters. The manufacturers only had one prototype and so users were able to sit at the workstation and see how it felt, but they could not appreciate what the cluster effect would look like.

In another project, the architect produced some three-dimensional representations of the final building. The technique employed by the architect was a worm's eye view (i.e. the building as seen from underneath, looking up through the different floors). The client and other members of the design team were unfamiliar with this technique and so held the drawings at various different angles trying to work out what they were looking at. Finally they asked the architect and he explained how to read the drawings! These drawings looked impressive, but only confused the client. Consultants should concentrate on producing drawings that are informative to clients and not just other consultants.

Summary and self-assessment

Overall this chapter has stressed the importance of escaping 'flat land' and presenting information in ways that bring it to life for all of those involved.

Table 6.1 focuses on the main factors necessary to achieve this and provides a self-assessment tool for the reader to assess where effort is most urgently needed. The tool can be used for the reader's organisation in general, some part of it, or it could be used to assess a specific project and just those people involved.

The outcome of the analysis is the identification of those factors that are a high priority. An explanation of how the table is completed, and how the resulting information can be used, is given at the end of Chapter 2.

It should be remembered that the factors highlighted at this stage are in only one area. There are four others, which together make up the 'five-box' model. It may be worth waiting until similar analyses have been done for the other areas and a combined summary created as shown in Chapter 12. It may then be even easier to link together high priority factors with issues that are important and urgent for the firm.

Table 6.1
Summary assessment of appropriate visualisation techniques

Instructions
For each of the factors listed below, circle one number in each of the two scales provided to indicate the extent to which the factor is currently met and the effort you are currently making to address the factor. A key for each scale is provided below.

Met?
This scale is to assess the extent to which each factor is currently met:

1 = Not met at all
2 = Met to only a slight degree
3 = Met quite well
4 = Fully met

Effort?
This scale is to assess the effort you are currently making to address each factor:

1 = No effort at all
2 = A little effort
3 = Quite a lot of effort
4 = Considerable effort

Factors	Met?	Effort?
V1. Visualisation techniques should be employed to increase potential for shared understanding	1 2 3 4	1 2 3 4
V2. Visualisation techniques should be adequately resourced	1 2 3 4	1 2 3 4
V3. Visualisation techniques should be used effectively	1 2 3 4	1 2 3 4

Instructions (continued)
When you have done this for each of the factors, use the two scores to locate each factor, using its list number, in the grid provided to the right of this box.

Entries that fall in the black areas can be ignored. For the remainder, the lighter the area the more urgent it is for there to be a concerted effort on the factor concerned. In this way high-priority factors can be identified.

These high priority factors can be combined with similar results for the other 'five-box' areas in Table 12.1 on page 133.

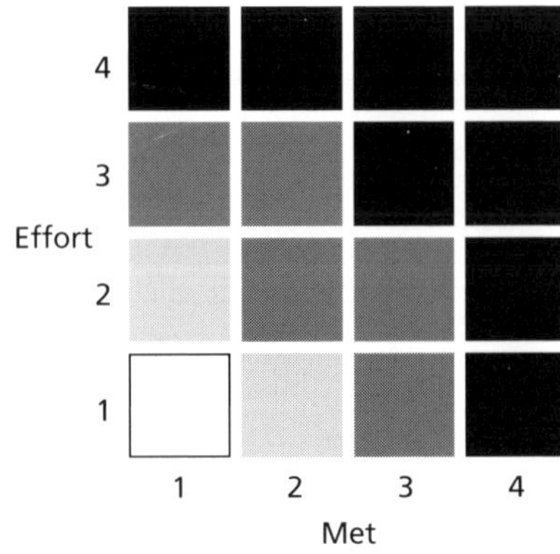

Summary assessment grid

7
Implementing change

7
Implementing change

Generally

It is one thing to have identified key improvement areas; it is another to achieve the improvement. There are many possibilities and the five identified areas give some indication of where a firm could choose to start. It has already been mentioned in Chapter 1 that the best practice advice over the last thirty years has produced little progress, and some considerable effort has been put into trying to understand why this is so.

It was also shown that proposals for change need to fit in with a firm's experience base, but two important features of successful implementation need to be explored further here. The first is that it appears that all of the five key areas outlined in Chapters 2–6 are intimately (systemically) connected, and so action in one area almost inevitably spills over into other areas in due course. The implication is that it does not matter too much where a firm starts, the effect is likely to be widespread, provided a sustained effort is maintained. Thus, a firm can choose to start in an area that has particular importance for them and about which they can get highly motivated. The second aspect is the rate at which improvement is achieved. It seems clear from our experience with companies introducing change that gains are only slowly accumulated. Those seeking to improve should be realistic and take an incremental approach, which, again, starts in one place and in due course makes significant progress, through sustained effort.

The common core of the above two aspects is the need to start somewhere and then to maintain the effort. The initial focus for the improvement, or indeed the specific initiative used to kick off the process, are both clearly necessary in terms of gaining commitment and enthusiasm from those involved. However, they are actually less important to the ultimate improvement gained than might be thought.

The interactive nature of initiatives

Table 7.1 shows, in summary form, the experience of four initiatives to improve briefing practice. It can be seen that in each case there was a primary focus for the initiative and that knock-on effects quickly accumulated in the other areas as well. In Project 1, the impact is widespread with every area becoming involved after an initial focus on empowering the client. Project 2 starts with user involvement, which then automatically has an impact on the empowerment of the client and quite naturally draws upon the use of appropriate visualisation techniques. In Project 3, the focus on project dynamics led to actions on appropriate team building. All of these examples reflect the main connections, shown in Figure 1.11, between the five key improvement areas. Project 4 illustrates an exception, with an initial focus on user involvement, but knock-on effects on project dynamics and team building. These last two are not suggested as primary connections but highlight the fact that in reality all areas interact with all other areas.

Table 7.1
Summary details of four initiatives to improve briefing

Intervention	**Project 1** Construction of innovative new high-tech building for inexperienced client	**Project 2** Conversion of existing building to new multi-user health care use	**Project 3** Construction of new build health care facility with experienced client	**Project 4** Briefing for the development of software to support briefing
Empower clients	➊	②		
Project dynamics	②		➊	②
User involvement	②	➊		➊
Visualisation techniques	②	②		②
Team building	②		②	②

Key: ➊ = primary focus; ② = evolving secondary foci

The improvement process

In addition to identifying a focus for improvement, the process of improvement itself demands attention. From our observations, implementation is very far from being an event and in fact deviates rather significantly from even a linear process! Firms in reality are very tentative when taking up new ideas – for some very good reasons. They are each part of a network of participants in a project and this means that it is difficult for them to change radically on their own. To quote one participant, 'our clients do not want to be guinea pigs'. Thus, the process can be seen to go through various stages, the first of which is hardly perceptible. Figure 7.1 shows these stages, drawing directly from experience with companies seeking to improve briefing practice. The diagram shows time along the x-axis and briefing performance on the y-axis. Initially the firms take tacit action, that is, they do what they have always done, very often for unexpressed reasons based on experience.

As a result there are considerable restraining forces to change, signified by the large negative arrow, and only fairly minor driving forces for change, signified with the small positive arrow. Any driving forces probably derive from a

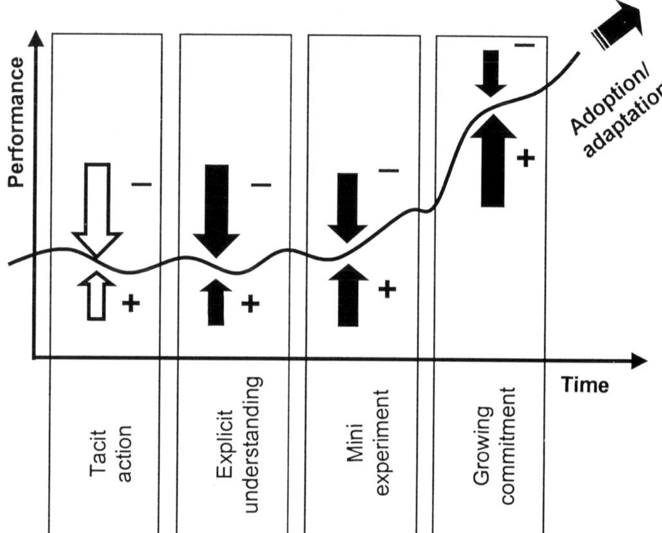

Figure 7.1
Practical improvement process

certain dissatisfaction with their performance compared with the perceived norms of the industry. On discussion, companies tended initially to resist implementing proposed good practice ideas. Given the relatively well-founded nature of the proposals this might seem irrational, and probably is so from the perspective of the industry as a whole. However, from the point of view of any individual player within the industry (with the possible exception of a powerful client) proposals for change have to be judged at a local level.

For many designers, for example, there is simply neither the time nor the resources to follow perceived 'good practice' and design a scheme in all its detail before going to tender. In addition, as the client is often in a hurry to progress, it simply may not be advisable to have extensive debates at a conceptual level. It may be more rational (at the local level) to stick to very pragmatic, tangible things, like room sizes, and let the more complex issues arise somewhat later. At a more general level, the individual players are trapped in a dynamic system in which their role is defined by interaction with many other players. To change independently of the rest of the system is potentially very high risk and *not a rational choice for a firm that wants to survive and succeed.* Thus, any changes must either be by consensus throughout the system of organisations involved – maybe by the client taking a strong lead – or they must be limited by the constraints of satisfying the existing relationships, whilst making progress towards desired improvement.

There may not have been much evidence of significant change, but a change had occurred in terms of the *understanding* that the firms had of the problems they were encountering. Although many of the same things still went wrong in the same ways, the reasons were much clearer and their occurrence was much less acceptable now that those reasons were explicitly understood. In fact when problems arose that triggered this knowledge afresh it often led to an 'if only we had...' type of reaction. This rising level of awareness led on to the next stage, which was for the firm to be sufficiently motivated to engage in a *mini-experiment.* This was a tentative, limited change that was not going to greatly impact on other members of the project, but had the potential to provide some benefit. Where

the mini-experiments had positive outcomes a *growing commitment* was observed as more of the proposals for change were adopted, but adopted in ways that fitted in with the existing experience and characteristics of the companies in question.

This step-by-step process interestingly reflected received wisdom on how people learn from experience by passing round a cycle (see Figure 5.2). So, after a slow start, the changes gained momentum. Sometimes it was too late for much of the benefit to accrue to the project being studied, but taking a broader perspective, the ideas could be seen to be taking root in the organisations in general. Thus, there was a transition from general improvement ideas to specific local actions that were owned by the companies involved. The emphasis is on 'user pull' and 'local context', rather than the imposition of generic 'best practice'.

The juggling analogy

Gelb, M.J. and Buzan, T. (1994) *Lessons from the Art of Juggling*, Aurum Press, London.

There is a striking resemblance between the pattern of adoption shown in Figure 7.1 and that predicted by Gelb and Buzan (1994) through their juggling analogy for managerial improvement. Learning to juggle is a matter of mastering a rather complex process where, when you concentrate on one aspect, you tend to make mistakes on some other; however, to appear at all credible you have to perform all of the elements together. This illustrates many of the tensions we have been discussing above. Gelb and Buzan go through the main features of a positive process to develop the ability to juggle. The key features are:

- Have a clear vision of where you are heading
  - For juggling maintain a 'soft focus' straight ahead
- Focus on a key facet
  - For juggling this is the throw, not the catch
- Encourage 'mistakes' so that learning can take place
  - Start juggling with one ball – and don't 'let' yourself catch it
- Build up complexity in stages
  - For juggling this requires 'relaxed concentration'

In more detail, a participant in construction needs to keep in mind the overall objectives that they are trying to achieve so that these can infuse all of their actions. This isn't the specific objective you are straining to meet at a particular time, but the general, overall philosophy behind all of your efforts. For construction generally, a 'soft focus' on client satisfaction or, better still, delighting the client would seem appropriate. More specifically for briefing, it could be to infuse the whole construction project with the client's requirements. Within this overall orientation the next suggested feature is to focus on a key facet, rather than trying to do everything at once. From the earlier chapters of this book it should be evident that we would suggest selecting one of the five boxes shown in Figure 1.11 in order to identify an important but achievable aspect to act upon. This is the juggling equivalent of concentrating on the throw rather than the catch. However, the implication of allowing this focus is that 'mistakes' must be encouraged in order that learning can take place. So in juggling you don't allow yourself to catch the ball so that the throw can truly be concentrated upon. In briefing practice this is, of course, slightly complicated as mistakes may undermine the credibility of the firm or lead to litigation. Thus, safe environments must be created so that the necessary expertise can be built up without causing these sorts of problems. If, for instance, a particular information-eliciting approach was to be taken with clients, this could be tried as a dry run within the company through role playing, before being used with real clients. The last feature, namely building up complexity in stages, is implicit in much of the above. If you are to focus on key facets and learn through mistakes, then this is bound to be on a limited front initially. However, to fully address the improvement aspirations implicit in the overall vision, various aspects need to be pulled together and the enhanced 'performance' launched into the project domain. This cannot be done in a high tension panic, but must be delivered with relaxed concentration, in the knowledge that the necessary preparation and build-up has taken place.

focus on a key facet, rather than try to do everything at once

safe environments must be created so that the necessary expertise can be built up

There are very clear parallels between the above process and the reality of implementing improvements as observed

in relation to briefing. Figure 7.2 gives the expected profile for juggling improvement, which is decidedly non-linear. Part of the secret of success with the approach recommended is, however, to recognise this pattern and not to be thrown off course by the dips in performance. Rather, if these are seen as preludes to rapid improvement, this completely changes how they are responded to.

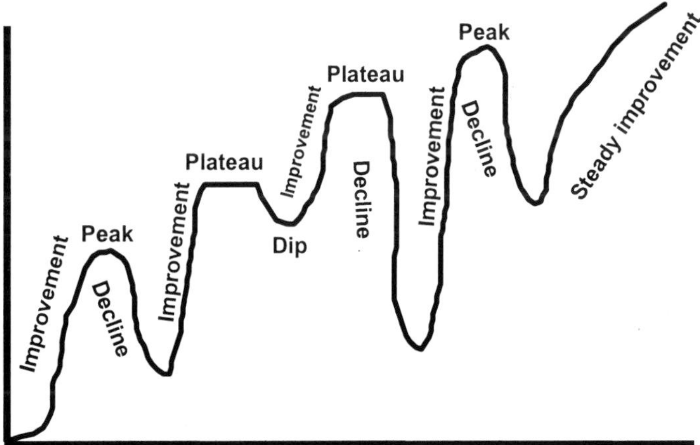

Figure 7.2
Improvement and progress over time

Summary

This chapter has tried to illuminate the implementation process that individuals and firms can expect to experience as they endeavour to use the briefing ideas presented elsewhere in this book. The next four chapters illustrate the experience of participants in four projects over a period of several months as they sought to use ideas focused around the five-box model. It should be pointed out that the whole design of this book and associated materials have been conditioned by an improved understanding of the implementation process. Thus, the thrust is not to provide 'good practice', but rather to illustrate key improvement areas, together with examples of alternative approaches, as far as possible set in their original context so that those considering them can better judge whether they suit their local needs.

This philosophy has been taken further in an experimental CD being developed in parallel with this book, in which users can search for illustrative examples in a problem-led way. An example screen from the CD is given in Figure 7.3 from which it can be seen that one way of searching for information is by project phase against key solution area. In addition, key word searches are possible and each line of enquiry ends with specific examples of actions taken, the outcomes achieved, and the context within which it all happened, based on real project experience. The CD is designed to support the incremental, non-linear implementation process that we expect construction participants to take. In addition, video material is planned which it is hoped will be especially helpful in raising general awareness, both within construction consultancies and client organisations.

For details of the CD and video, contact the Research Centre for the Built and Human Environment, Bridgewater Building, University of Salford, Salford, M5 4WT, UK.

the incremental, non-linear implementation process that we expect construction participants to take

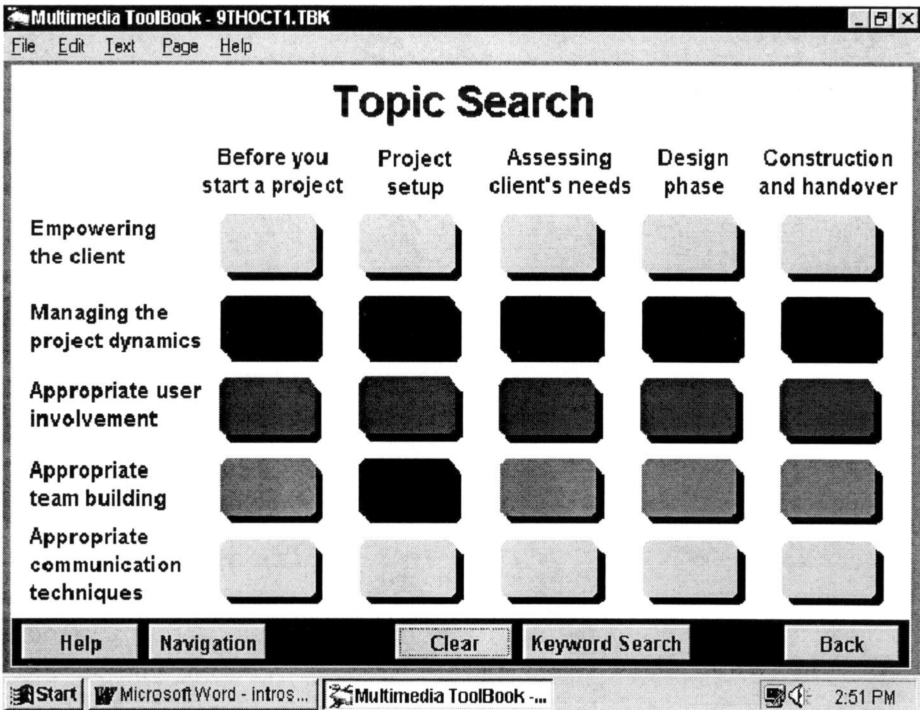

Figure 7.3
Example CD screen

In summary this moves the argument on from why 'good practice' advice does not appear to work, to a better understanding of a realistic implementation process that acknowledges existing experience, local conditions and the effect of multiple participants in the activities of any one player within a construction project. An overall summary of the sort of process that people using the suggestions in this book should expect to engage in is given in Figure 7.4. This is a development of Figure 7.1 in that it stresses the tentative take-up phases as before, but it now also includes a clear indication of a continuous improvement process (upwards spiral) thereafter, as the systemic connections take effect and confidence grows.

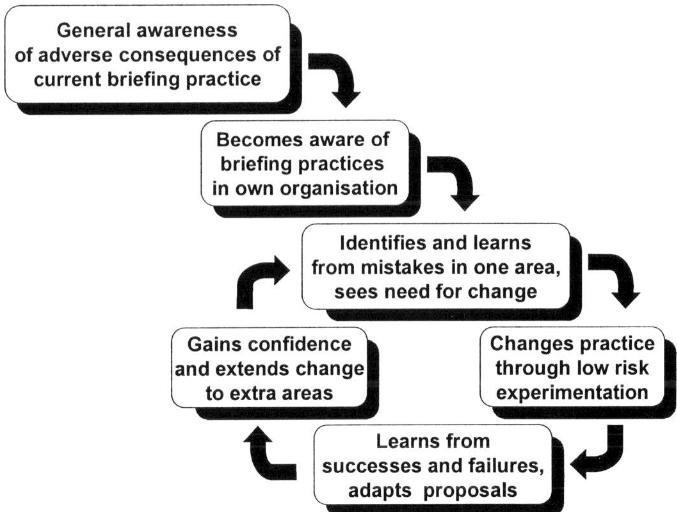

Figure 7.4
Continuous implementation process

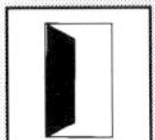

8–11
Case studies 1–4

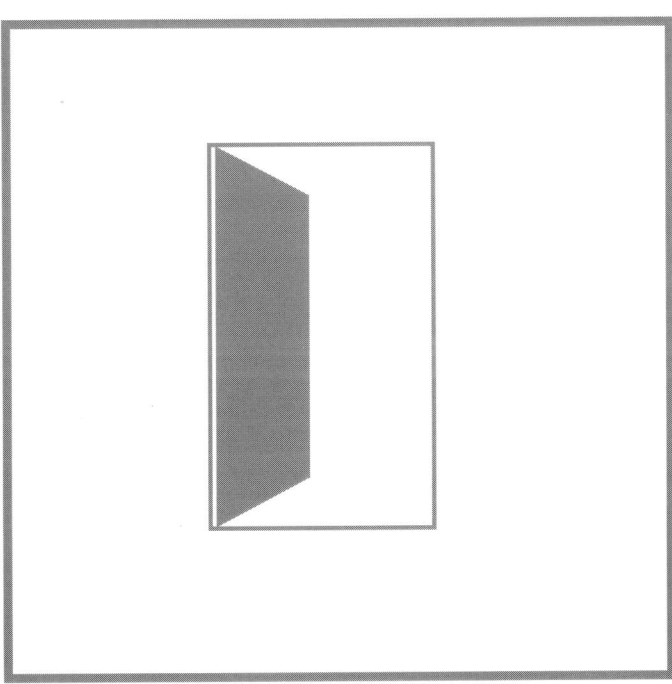

8
Case study 1 – Empowering the client

Introduction to case studies

This chapter, along with Chapters 9–11, illustrates the experience of participants in four real-life projects over a period of several months as they sought to use ideas focused around the five-box model.

Notes: In all of the projects it was agreed with the project participants that one researcher, maybe two, would observe the projects as they progressed. In each case, after initial discussions, the researchers suggested an area based on the five-box model where an alternative approach could prove beneficial. However, the project participants were free to decide whether or not they would follow this advice.

The main body of the text describes the development of the project. The commentary in the margins highlights that the five key improvement areas are closely connected and that effort (or lack of it) in one area will have an impact in another area.

Overview of project

The inspiration and driving force of this building is a recently developed technology. The building is a joint venture between two organisations that share an interest in this specific technology and wish to promote its potential applications. One is commercial (Client A) and one is a research group (Client B). Client A will own the final building and Client B will sub-lease part of it. Both organisations will have dedicated areas, but some facilities will be shared, and funded, by a joint company that they are setting up. Client A is represented by a project manager, plus a senior manager who will attend some meetings. Client B is represented by the head of the research unit, plus a business manager.

Focus of main intervention: empowering the client

Client B, the research unit, are experts in the technology, but they have had no previous contact with construction and so are not sure what is expected of them during the construction process. It is agreed they might find it useful if the researchers (see margin opposite) attend project meetings with them and assist them as necessary to brief the architect. In essence the researchers will act as an 'expert' client ensuring that the actual clients have thought about their requirements and conveyed them properly to the rest of the design team.

Influence of other factors

In this case even though the primary focus is on 'empowering the client', it will become obvious that 'managing the project dynamics' and 'appropriate team building' actually become the critical factors that dominate the development of the project.

Implementation

Initially there appears to be resistance from Client B when the researchers make suggestions, even though the client is inexperienced. The client seems to feel that the researchers are more into theory than practice and that the 'real world' requires immediate action, not thought or reflection. As the project develops this attitude changes as the client begins to realise that they really do need to think about their functional requirements in much greater detail.

Project development

Meeting 1

Client B (the naïve client) meets with the researchers to discuss the client's responsibilities during the early stages of the project. The client realises that they will need to brief the architect/design team on their requirements and so they have produced a draft 'functional specification'. This specification is a useful starting point, but is far more technical than it needs to be at such an early stage. The client has suggested already which rooms will be required and for each space there is a list of equipment. There is no mention of

Naïve clients are unlikely to know what information the design team will require and so design teams should brief the client on what sort of information they need. The design team will obviously need to know what will occur in the building, but it will also

help them to know why the client needs the building and what the organisation's aspirations for the future are. The better the design team understands the client organisation, the more likely they are to design a building that really supports the client's aims. Thus the design team also becomes empowered.

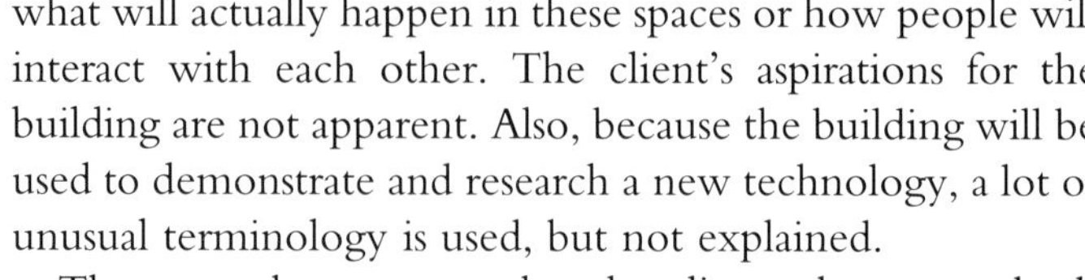

what will actually happen in these spaces or how people will interact with each other. The client's aspirations for the building are not apparent. Also, because the building will be used to demonstrate and research a new technology, a lot of unusual terminology is used, but not explained.

The researchers suggest that the client takes a step back and tries to think about why the building is necessary and what they are trying to achieve with it. They should consider what activities will take place, rather than try to specify what rooms are needed. The client is obviously not convinced that this is necessary and says that they are working in the real world and so need to fix the sizes of the rooms as quickly as possible as they are working to tight deadlines.

Meeting 2

The project team did not consider in detail how the project would be managed. They moved onto room sizes without really agreeing what the aims of the project were. This meant that time was wasted later when this subject had to be revisited.

At the first meeting with the architect the discussion moves straight onto room sizes. Client B has supposedly produced a revised brief, but upon inspection this is a matrix showing the previous information, i.e. rooms with lists of equipment. It is still impossible to tell what will happen in each room and who will occupy it. The architect has not met with Client B before and is not really given time to ask them any questions. Client A seems keen to agree upon room sizes as quickly as possible.

Meeting 3

This intensive questioning and detailed examination of each area demonstrated to the client that there is far more to briefing than just supplying a list of rooms. They could see that unless the design team can obtain a vision of what the client wants to achieve operationally, then it is impossible to design an appropriate solution.

At the next meeting the architect hands out a draft brief that lists the different rooms, their proposed sizes and a bubble diagram which indicates the relationships between the areas. It is decided that each room will be considered in turn to see if it is the right size. As the meeting progresses it becomes clear that there is still a lot of confusion over what will actually happen in each area. The researcher decides to ask a lot of operational questions to try to improve everyone's understanding of the situation. For example, Client B wants a quiet room away from the main offices where people can write-up reports. No-one else thinks this is necessary, as people will have their own workspaces. The client explains that the computers at the workspaces are extremely specialist and won't actually be used for simple wordprocessing. This is not obvious from any of the briefing material and is only highlighted by intensive questioning. Following on from this, Client B draws each area out on a white board, explains what sort of person will occupy each space and what they will want to do in each area. This gives everyone a much better appreciation of what will happen in the building.

This meeting also serves to highlight communication and understanding problems. Client B is an expert in the technology, but finds it difficult to explain what is required without using jargon that is unfamiliar to everyone else. Client A says that everyone is 'wrestling with a different vocabulary that we don't share'. Similarly Client B does not always understand the construction terminology used by everyone else. Also part of the building will be used to teach the technology and it emerges that the architect is unfamiliar with teaching practice and is making incorrect assumptions. The researcher feels compelled to ask questions concerning teaching methods to highlight any specific functional requirements so that the architect doesn't overlook anything important.

Design team members do not necessarily need to be familiar with specific building types as experienced consultants should be able to interrogate a client's brief so that everyone gains a shared understanding of the situation. However, inexperienced consultants should be trained by senior staff to ensure that they ask appropriate questions.

The meeting also shows that there are communication problems between the representatives from Client B. They have not discussed, or agreed, their requirements amongst themselves prior to meeting with the rest of the design team. So quite a lot of the meeting is spent clarifying what the different members of the client team actually want and how they are going to reconcile their different requirements. Such discussions are not really of interest to the design team and people seem to be slightly annoyed that their time is being wasted.

Client teams should discuss their objectives and needs before design meetings to ensure that they have the same aims. Also hasty decisions made at meetings without considering the implications are likely to result in changes later.

Even though all parties seem keen to agree on room sizes, Client B appears to find it difficult to translate dimensions into reality. Time and time again during the meeting it is necessary to try to explain dimensions to the client by making comparisons with the room that the meeting is being held in.

Architects should take tape measures to meetings so that they can properly demonstrate room sizes.

Visit to relevant building

Visits to similar buildings are a good way for the client and design team to try to obtain a common understanding of what sort of solution the client is seeking. By having something specific to look at clients can more easily demonstrate their likes and dislikes. It may be possible to 'borrow' an existing solution and use it in the proposed building.

As the design team are finding it difficult to understand the technology as described by the clients, it is agreed that a visit to another building where the technology is in place may be a good idea. It is proposed that Clients A and B, design team members and the researchers all visit together. Unfortunately the clients visit the building on a separate occasion from the rest of the design team. When the design team and researchers visit the building they are not able to question the clients directly concerning how the technology will impact upon the proposed building and so the visit is not as useful as it could have been. However, the chance to view the equipment and discuss the implications (even at a late stage) still makes a vast difference to everyone's understanding.

Later meetings

Where possible the consultant representatives should remain the same throughout a project. If new people attend and suggest alternative approaches the project may be held back. In this case when a senior manager attends a meeting and asks questions the whole team (and project progress) is thrown into turmoil.

As the project progresses, the design team members do not remain constant, as different representatives are sent from the involved organisations. These people have not always been properly briefed and so the scheme fundamentals have to be explained at each meeting. Obviously each new person has a different perspective and so the scheme keeps on changing as new ideas are added.

At one meeting a senior manager from Client A questions why the building is the shape that it is. The shape is one of the few fixed parameters as far as the rest of the design team are concerned. They are noticeably annoyed that such a fundamental parameter does not appear to have the full backing of Client A and could potentially be altered at such a late stage in the design process.

If clients continue to participate fully in the design process, rather than just the initial briefing, the completed building should suit their needs as they can consider the various implications on their business as the design develops.

As the project develops the researchers take on the role of observer as Client B's representatives become more familiar with the construction jargon and are happy to contribute fully in discussions. They now realise that as clients they need to play a continued part in the design process if they want a building that really meets their requirements. They are the only people who can assess if a specific design decision will enhance or detract from the way that they will utilise the building.

A number of outline schemes are proposed by the architect, but the schemes do not meet fully with the approval of both clients. It appears on a number of occasions that the architect has been in consultation with Client A who has made decisions on behalf of Client B. These decisions relate to Client B's spaces, not the shared ones, and so Client B is forced to ask for changes during meetings which results in abortive work. Client B is annoyed that the consultants are more concerned with meeting Client A's needs than Client B's.

Projects with more than one client are always complex. It is imperative that all involved parties are aware of the specific relationship between the clients and that everyone understands (and agrees) how the project will be managed. In this case, Client A will own the building and has employed the consultants. Hence the consultants appear to find it necessary to put Client A's preferences before Client B's.

As the project progresses Client B becomes more frustrated that the project is not being managed very well and complains that no-one appears to be making any firm decisions. They question how buildings are ever finished if this is normal practice. Client A's project manager is supposedly pushing the scheme forward, but whenever he is asked a question he always says that he will have to refer to senior management before he can make a decision. New sketch schemes are presented at each meeting and then by the next meeting the scheme has totally altered for no apparent reason.

Finally an outline scheme is proposed which appears to meet with everyone's approval. Client B clearly states at the meeting that they are happy that the scheme meets their requirements. Client B leaves the meeting assuming that the design team will now pursue this scheme in detail. However, at the next meeting the approved scheme is nowhere in sight and the architect presents a totally new scheme. Client B is amazed that this is occurring and asks why their comments have been ignored. It is suggested that the previous scheme was over budget. However, certain members of the design team claim that the new scheme is more expensive as it is more complex structurally. It seems likely that there are other driving forces behind this change in decision. Even though Client A's project manager approved of the original scheme, this change indicates that the scheme was overturned by Client A senior management. Client B leaves the meeting very annoyed that they seem to have no control over how the building will develop.

This change of heart demonstrates once again that Client A's project manager has no real authority to make decisions and that senior management are able to overturn major decisions without reference to Client B. If design team members are not authorised to make decisions and they can be overruled, the rest of the project team can never be sure that the project is progressing on a sound footing.

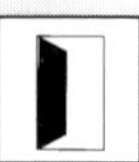

Where possible it makes sense to employ consultants who could affect the design as soon as possible. Other consultants will not want to carry out work which turns out to be abortive.

At a later meeting a new consultant is introduced by Client A to examine opportunities for commercial sponsorship of the building. The consultant says that sponsors will want input into the design. The design team are once again visibly dismayed that such an influential decision has been delayed and that they could possibly have to start redesigning all over again with a new set of parameters.

Summary

In this case Client B had no previous experience of construction projects and so the focus was on 'empowering the client'. At the start of the project, Client B was of the opinion that all they needed to do was pass on a list of their requirements, with room sizes, and that the construction team would magically design an appropriate building. They soon realised that if they did not participate fully in the process and really explain why a particular function was needed, the final building would not be suitable for their needs. Initially Client B did not contribute to project discussions as the terminology and procedures were unfamiliar. However, as the project progressed the client became more knowledgeable and confident and was soon asking pertinent questions of the design team. It is unfortunate that as the client became more empowered they also became more disillusioned with the process as their requirements appeared to be secondary to those of the other client.

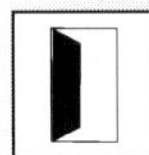

9
Case study 2 – Appropriate user involvement

Overview of project

This case study concerns the refurbishment of an existing building. The building has not been used to full capacity for several years as it needs a total redesign to bring it up to date. Some members of the design team have worked previously with the client and all have experience in the relevant field. The client has appointed two internal project managers to coordinate the client side of the project. However, neither of them has had any previous construction experience and both carry on with their normal duties as the project develops. An external project manager has also been appointed. A four-stage briefing process has been planned over a four-month period. This comprises schedules of accommodation, block diagrams, room layouts and outline plans. User input and feedback will be sought at each stage.

Focus of main intervention: appropriate user involvement

The main focus of this case study is 'appropriate user involvement'. The client had on two previous occasions suggested redeveloping the building, but nothing had happened and users had been disappointed. This time senior managers want to involve the users in the briefing process to prove that they are serious about refurbishing the building. A number of departments could potentially occupy the redeveloped building and so these departments are asked to select representatives to meet with the architect. The researchers want to see what impact the users will have on the project and so plan to attend meetings between the

architect and the different user groups. The researchers will also attend general project meetings to monitor progress.

Influence of other factors

Even though the focus in this case is 'appropriate user involvement', the way the project is managed becomes increasingly important, as does the lack of decision-making within the client organisation.

Implementation

This project in particular highlights the difficulties of trying to get organisations or individuals to implement new ideas. The researchers had gained access to the project via the project manager, unlike the other projects where the client had been the main contact. On several occasions during the briefing process the researchers suggest that alternative practices may help to solve certain problems. On each occasion members of the design team are supportive of these suggestions, but the client blocks them or only allows minimal take-up. This case study demonstrates that without client backing it may be very difficult to introduce alternative practices.

Project development

Meetings between architect and user groups

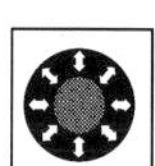

The number of users who attend meetings with the architects, etc. should be restricted. Large groups make decision making very difficult. It is better if users select representatives to attend and brief them properly on what is required.

The client's project managers do not appear to have placed any controls over who will attend user group meetings and so the departments are free to send whoever they like. This means that the number of user representatives for each department varies, and in one case there are over 20 attendees. Most groups appear to be evenly represented by senior managers and other members of staff. The lack of client direction concerning the user groups causes several problems which are apparent from the outset.

When the architect meets with each group for the first time, he begins by explaining what is expected of the users. The architect outlines the timescales, what the briefing process involves and what deliverables are expected. This seems like a useful introduction and the researchers assume that the users will now understand what they are expected to do. Subsequent meetings demonstrate that the users do not appear to grasp what they should be doing and the architect has to frequently point out their responsibilities. The client's project managers are actually responsible for guiding the users through the process and for ensuring that they meet deadlines. However, their lack of construction experience means that they are not sure how to manage the user groups properly.

It is important that users are briefed well on their responsibilities and what they are expected to produce. If they do not understand the importance of the programme, for example, the project will probably be delayed.

Users find it difficult to allocate time to attending user group meetings; quite often people have to leave before the end of meetings, as they have to return to work. Even though the client wants the users to be involved, the client's project managers, or senior managers within each department, are not specifically allocating time for users to participate in the project. This sends a message out to the users that the client does not actually think the project is very important, hence the users do not take it seriously either.

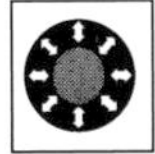

If clients want users to be involved in the process they need to ensure that the users can take time out from their regular work to consider the project in appropriate detail.

The purpose of the user group meetings is to discuss departmental requirements with the architect. Few groups have discussed and agreed what they need prior to the meeting, so there are many discrepancies. Users cannot decide how their departments should function and what spaces they need. The architect spends many hours listening to internal disputes which should really have been sorted out away from these meetings. There is also an increased risk that the users will change their requirements again once they have had time to think the situation through properly.

Users should agree between themselves what their requirements are before their representatives meet with the architect, etc. They should try to prioritise their requirements as it is likely that some concessions will have to be made.

The mix of user group members creates further complications. Some groups are dominated by senior managers and people seem afraid to speak out against them. On one occasion a senior manager makes a decision which everyone agrees to, and when he leaves, the other users try to overturn it immediately saying that he doesn't really understand what is needed.

When user groups are formed, the mix of personalities and hierarchies should be considered. No one person or subset should be allowed to dominate.

User group members should remain constant if possible. If new people attend, the original requirements may well be forgotten as people often focus on their own priorities, rather than the group's.

Many users will have no previous experience of construction projects. Such users need explicit guidance on what decisions they are expected to make, or else they will waste time on irrelevant issues. Written instructions or standard pro formas may be useful if there are several groups as each group will then be working from the same starting point. Depending upon the resources available, it may also be a good idea to employ a facilitator who understands the business and what is required by the construction team.

Each department meets with the architect on several occasions to verify that their requirements have been interpreted correctly. As the members of the user groups are not fixed, different people attend each meeting. This means that additional requirements are introduced and people query past decisions.

As the meetings progress the architect becomes more frustrated with the situation, as the users' inability to meet deadlines is slowing the project down. The users are also complaining that they still don't really understand what they are supposed to be doing. The researchers and the project manager discuss how the situation can be improved. It is proposed that the researchers should spend time with each group away from the formal meetings, acting as facilitators to encourage the groups to make their own decisions and prioritise their requirements. The researchers will also try to aid users' understanding of the architect's proposals. These suggestions are welcomed by the architect, but the client is not keen to let the researchers have separate access to the users. It is suggested that the researchers may upset the political situation and encourage the users to make further space demands. At a later date the client agrees that users can approach the researchers if they are having problems, but this is a half-hearted attempt to sort the problem out as it puts the onus on the already overworked users to make contact. By this stage most users have become disillusioned with the process and have stopped attending meetings anyway. Unsurprisingly, few users consult with the researchers.

The situation regarding the users does not improve and the programme falls further behind schedule. Finally at one meeting the architect states that the block diagrams are nearly complete and only need fine-tuning. He states that from now on he would appreciate it if as few users as possible from each department are involved. The room data sheets are to be issued to selected users identified by the client's project managers.

Maintaining user involvement

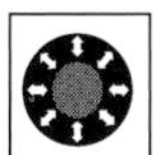

If clients wish to involve users in the briefing process they need to convince the users that their opinions count and that their requests will be taken into consideration. If clients only involve users because it is seen as the thing to do, but do not actually intend to let the users have any say, then it is pointless and dangerous to involve users at all. Users who are consulted and then ignored will feel aggrieved. In this particular case, the chief executive is giving mixed messages; he wants to involve the users, but would rather they attended progress meetings after work so that their everyday duties are not affected.

The users have twice previously been told that the building was to be refurbished, only for nothing to happen. They are naturally sceptical that anything will occur. The external project manager is keen to demonstrate to the users that this time something will happen. He proposes that a series of user liaison meetings be held where the users will be informed of project progress and where they can ask questions. The researchers fully support this proposal as the users in the case studies had frequently complained that after passing on their requirements they heard nothing more about projects until they occupied them. The chief executive is approached and supports the idea. He proposes that these meetings should be held at 5:30 in the evening so that people can attend after work. The design team and researchers feel that the timing immediately indicates to the users that the project is not viewed as very important by senior management as people have to attend in their own time.

The client's project managers are supposed to explain to the users that these meetings are to keep them informed of project progress. However, at the first meeting the users take the opportunity to try to question the chief executive about which departments will occupy the building and why. The chief executive will not answer any questions concerning the space allocation. The users complain after the meeting that they have no real influence over the direction of the project. The next meeting follows a similar pattern with users asking why they are not going to be allowed in the building. It is decided to suspend the meetings as the users and client obviously have different agendas and the meetings serve no purpose until concrete decisions concerning occupancy have been made.

Client management of project

A client's main responsibility during a construction project is to take decisions and choose between alternatives, so

As the project progresses many users become disillusioned with the project, and fewer people attend their respective user group meetings. One reason for this may be the reluctance of the client to make a decision concerning who will actually occupy the building. It was decided at the outset

that progress can be made. This client delays making decisions for various internal political reasons. This affects project progress and upsets the users.

that specific departments would put their proposals forward, but there was no guarantee that they would all be moved. After months of briefing, the situation has still not been resolved. Some of the users feel that they are wasting their time attending meetings when it is possible that they will remain where they are anyway.

Visits to other buildings

During the case studies the researchers had found that visits to other buildings stimulated innovative ideas that would probably otherwise not have been considered. Design time was also saved as designers were not reinventing the wheel when an appropriate solution already existed.

Some of the users mention other facilities that they have either seen or heard about. They wonder if parts of these schemes can be included in the design. The researchers suggest to the design team and client that it may be a good idea if the design team and/or users visit these establishments to see if they have features that can be incorporated. The design team is supportive of this suggestion, but for reasons unknown to the researchers these visits are never arranged. At a later stage it is suggested that the researchers visit the other establishments and report back whether it is worthwhile other people visiting. The researchers do not feel able to do this as they are not experts in the field and would not be able to appreciate which aspects were appropriate or not. The researchers believe that this is a wasted opportunity, but suspect that the client is not keen to spend money or time allowing users or the design team to visit.

Visualisation

When mock-ups were utilised in the case studies the users were able to see exactly what the architect was proposing. The users were also able to demonstrate how they would work in the space; often this differed from what the architect had imagined, and so the design team also learnt from the experience.

After the users have passed on their requirements, the architect works up some departmental layout drawings. Many of the users find it difficult to understand exactly what is being proposed and cannot easily imagine the sizes involved. The architect discusses the problem with the researchers and they all decide that it would be useful to mock-up some of the more important rooms to give the users a feel for the proposals. There is plenty of available space in the actual building that will be refurbished, as much of it has been out of use for a while. These suggestions are put to the client's project managers for them to discuss with senior management. Again, the architect never receives the go-ahead.

Summary

At the outset of this project it appeared that the client was keen to involve the users in the briefing process. However, as the project progressed it became apparent that the decision to involve the users was not being taken seriously by the client. Users were expected to slot their project duties into their already busy schedules and attend some meetings in their own time. Consequently, the users felt that the project was not seen as important by senior managers. In addition the users never really understood what they were expected to do.

The user contribution to the project really needed to be more tightly controlled by the client's project managers, but they too were unsure of their responsibilities. In addition, whenever major decisions needed to be made, the client's project managers did not have the authority to do so. The project would have run much more smoothly if the client had spent more time working out how the project would be managed from their side.

The client also failed to support the design team's efforts to increase user understanding of the project. Several techniques were proposed, but none of these suggestions were ever taken up by the client. However, these proposals were viewed as exemplary ideas by the design team and so it is likely that they will try to incorporate them in future projects.

10
Case study 3 – Managing the project dynamics

Overview of project

This case study considers a new building. The client has built several similar buildings recently and has its own in-house construction project management team, so can be considered an expert client. One of the in-house project managers would be responsible for the management of this new project. The other members of the design team comprise consultants who have previously worked with the organisation on other projects.

For this project the client organisation wanted to try to speed up the project time and so decided that they would probably adopt a construction management procurement route. With this approach the project is divided into different packages of work which are designed and tendered for separately. This means that construction can begin before all of the detailed design work has been completed and thus the project time can be reduced. This differs from traditional procurement methods where the whole design is completed, put out to tender and then construction begins.

Focus of main intervention: managing the project dynamics

It was suggested that the alternative procurement method would probably affect the way the brief was handled as certain decisions concerning the detailed design of the building could be delayed. Thus it was agreed that the focus of this project would be 'managing the project dynamics'. The researchers would consider what implications the alternative approach had on the management of the project;

what decisions could be delayed and what decisions need to be made early on in the process.

Influence of other factors

'Appropriate team building' features heavily in this case as the different members of the design team try to adjust to the new approach. Users are only marginally involved in this project as new staff will be taken on when the building is complete, but the lack of meaningful staff input highlights the importance of 'appropriate user involvement'.

Implementation

This project differs from the previous ones in that this time the client is responsible for the introduction of the alternative methods, rather than the researchers. This means that the rest of the design team really have to comply with the suggestions if they wish to remain on the project. In the other projects it was often difficult for the researchers to implement new ideas as they did not have the support of all involved parties. However, even on this project the client finds it difficult to gain the support of the rest of the team and to get them to change their practices accordingly.

Project development

Project team selection

Many experienced clients find it beneficial to employ the same consultants as they do not spend time getting to know each other's working practices, so they should begin to work productively from the start.

The client organisation has built several new buildings in recent years and has refurbished many of its existing buildings. During this time the organisation has employed a number of construction consultants on a regular basis. They have found it beneficial to employ the same people as the consultants have become familiar with the practices of the organisation and so do not need a period of familiarisation. One of the client's main aims on this project is to try to reduce the project time and so the organisation has

employed a group of consultants who have recently worked together on another of the organisation's projects and so are well up to speed. In addition, the representatives are all senior members of staff within their practices and so have the authority to make decisions on behalf of their organisations.

Early meetings

People are often very attached to their traditional methods of working and are wary of alternative approaches. Even though these consultants have worked together before, a re-evaluation of their working practices and relationships will be necessary in the light of this new approach.

Even though the consultants have recently worked on projects for the same client, it is apparent from the initial project meetings that all of the consultants, apart from one, are unfamiliar with the proposed construction management procurement method. They have a number of reservations about using this method. The consultants question whether this approach has been used on such complex buildings previously. They are unclear about their responsibilities under such a contract. How does it differ from JCT80, for example? What are their roles when the project moves to site? Will an increased number of tender packages increase the amount of co-ordination required?

In order to try to clear up some of these questions, draft contracts outlining roles and responsibilities are produced by the quantity surveyor and client's project manager. The contracts are discussed at length over a number of project meetings and reworded as appropriate. These discussions seem to take precedence over the project design for a while, highlighting the level of discomfort amongst the project team. The contracts remain in draft form for several months as the client organisation's board needs to approve the project and procurement method before the contracts can be finished. This means that the design team are working on trust for months before they sign the contracts.

The client in this case is experienced, knows what is wanted and is very much in control of the management of the project. This affects the balance of power amongst the team members. It appears that the consultants disagree

After a few meetings, initial project costings are undertaken and the project is over budget even at this early stage. The client's project manager suggests potential areas where savings could be made. The other consultants are not convinced that the proposed savings could actually be achieved, but agree begrudgingly that it may be possible. This discussion concerning costs highlights team structure complications. The client's role in this project is different to that of the traditional client as the client's project manager is actually performing dual roles. Not only is he providing briefing information, he is also effectively managing the project. The project manager takes advice from the other consultants as would normally happen, but does not always agree with their suggestions and so overrules them. On one occasion, the client's project manager and the quantity surveyor have a heated discussion about possible rises in inflation. Even though the quantity surveyor is supposedly more knowledgeable in this area, the project manager does not take his advice. The client is obviously paying for the work and so the consultants really have no choice but to follow the wishes of the client's project manager, even if they do not agree.

with the client, but they are forced to comply, even though they think the targets are unachievable. If the client was less experienced the consultants may well have had a different attitude.

Decision making

Decision making, or lack of it, is one of the major criticisms throughout this project. The consultants suggest that the client has not always made decisions at an appropriate time. For example, one of the work packages relates to the structure of the building. There is much discussion concerning the relevant benefits/disadvantages of steel versus concrete. Initially the speed of the project is thought to be of major importance and so steel is the ideal choice. Then it appears that the cost of the building needs to be reduced and concrete is a cheaper alternative. Naturally this decision has major implications for the rest of the design and the design team are anxious for a decision to be taken. However, the client seems reluctant to make a firm decision concerning the structure. Similarly, even though construction management is the suggested procurement route, no firm decisions are made and several months into the contract the client's project manager is still saying, 'If we follow the construction management route . . .'. The consultants appear to be unhappy that such major decisions are left to slide.

One of the most important factors to be taken into consideration during construction projects is decision making. It is only when decisions have been taken that the project can move forward. In this case the client wants to delay decision making as long as possible, although the reasons for this are not really understood. The client wanted to complete this scheme as quickly as possible; however, they are effectively delaying the scheme. The consultants find it difficult to design the project when such major decisions, such as the structure of the building, are being ignored.

One of the reasons why the client's project manager is unable to make decisions is that the organisation requires the project to gain board approval before it can progress to the next stage. The organisation has set procedures and deadlines for when the project can be presented. This internal set of procedures is not explained properly to the consultants,

Consultants need to be made aware of clients' internal decision-making procedures, so that they can synchronise the progress of the project with meeting cycles, etc.

This withholding of information indicates that even though the consultants and the client are supposed to be working towards the same goal, there is still a certain amount of mistrust between the two sides.

hence their frustration at the lack of progress. They believe that the project manager is responsible for decision making when really he cannot make decisions without reference to senior management.

Information control

After various attempts to reduce the project costs, the project is still £90 000 over budget. The client's project manager does not address this problem in detail, which is surprising as he has been pushing for savings at previous meetings. However, after the meeting the project manager informs the researchers that the organisation budget allowance for the project is actually greater than he has told the consultants. He says that if the consultants were informed that more money was available they would probably find ways of spending even more! This also explains his previous attitude to inflation as he knew that additional funds existed.

During the same meeting there are indications that information is not being distributed properly through the team. Some consultants seem to have more up-to-date drawings than others and so everyone is not working on the same basis. There are also complaints that the flow of information is not as good as it could be. One of the consultants blames another team member for not passing on relevant information. Due to the complex nature of the building much of the consultants' work overlaps – one member cannot progress until he has received the latest information from another. The client's project manager suggests that the construction management method may require an alternative approach to normal and that information should be passed on in sketch format so that the scheme can progress as quickly as possible. As the consultants already feel that the client is not providing their information at the right time the friction between the different parties continues.

User input

The new building will not be completed for a couple of years and so the organisation has only identified which functions will occupy the building and not the specific people. This means that the design team are unable to consult with the users of the building. Initially the lack of users is not a particular problem as there are various appropriate design guidelines that the team can refer to. Also all of the members of the design team are very experienced in the area and so can make reasonable assumptions.

The client's project manager also consults existing staff within the organisation who act as substitute users. Their preferences may not be identical to the final users, but as the organisation owns over thirty similar buildings it is likely that they can suggest typical solutions. As the project progresses, however, it becomes obvious that the design team really need detailed input concerning the operational requirements of this specific building. There is discussion within the client organisation whether it is possible to appoint the manager who will eventually run the building. The organisation is sympathetic to this request but is not keen to appoint (and finance) someone who will act purely as an advisor until the building opens.

After several months the manager is appointed. Although the design team have tried to predict what users will require, the manager begins to request changes. The project design is quite well established by this stage and so the design team find it difficult and time-consuming to accommodate the alterations. They suggest that the manager really should have been appointed sooner as he is the only person who could have realistically provided the relevant operational knowledge.

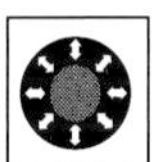

This refusal to appoint a manager demonstrates that the organisation does not really understand the importance or impact of user involvement. As the manager is employed at such a late stage the design team is forced to make assumptions, many of which are changed by the manager. Thus much of the project has to be reworked, which wastes more time and money.

Summary

In this project the client was responsible for the introduction of the alternative approach. However, the consultants were wary of following a new method, of which they had no experience. Even the client does not appear to have total faith in the proposed method and delays confirmation that it will definitely be used. This demonstrates just how difficult it is for construction professionals to adopt alternative practices. All parties have to be committed to the change or else it will be very difficult to implement.

11
Case study 4 – Appropriate user involvement

Overview of project

This project differs from the previous examples as it does not actually focus on a construction scheme, but instead considers the development of a computer-based briefing/feasibility tool. The tool has been developed over several years by a multi-disciplinary construction organisation. During the initial development phase, the system has been used on a limited number of real construction projects by the main person responsible for its design. The system is now ready to be tested more widely.

Focus of main intervention: appropriate user involvement

Before the computer-based briefing/feasibility tool can be distributed throughout the organisation it needs to be tested rigorously for user-friendliness and accuracy of data. A number of in-house professionals have shown an interest in the development of the system and it is decided to test it out on them initially. It is agreed that the researcher will sit in on user testing sessions in order to monitor how easy users find the system and to provide feedback on training methods.

Influence of other factors

The computer-based briefing/feasibility tool, or any similar program, could potentially have an effect on the way that construction projects are managed and the way that the team members interact with each other. Hence it was agreed that the researcher would also try to find out from the users what the implications of such a system could be on 'managing the project dynamics' and 'appropriate team building'.

Implementation

The main difficulty encountered is getting people to spend time testing the system. On several occasions testing sessions have to be cancelled as the users have to work on real-life projects. Even though the development of the program has the general backing of senior managers, they do not actually seem to free up people's time so that they can test the system properly.

Project development

System overview

Briefly, the aim of the program is to provide the client/design team with enough costing information at the feasibility stage to allow them to see if a scheme is worth pursuing. In order to calculate these costs the system asks a series of detailed questions about the proposed scheme and the user has to input the relevant data. Thus, the project team has to make specific briefing decisions at a much earlier stage in the process than normal. For example, how will the building be constructed, how many floors, its layout, etc. The more information that can be provided, the more accurate the estimate is likely to be. The system allows alternative options to be input so that comparisons can be made.

User involvement

The researcher attends a number of testing sessions to ascertain how user-friendly the system is. A single user attends each testing session and is taken through the system by its developer. Even though the users in this case are testing a computer program it is interesting to note the similarities

between the problems encountered here and those faced by user groups in the construction projects.

The sessions are unstructured and consequently the users find it difficult to understand what the purpose of the system is or why they are needed. Are they testing for user-friendliness, accuracy of data or the design of the system? It is agreed that some sort of structure needs to be developed so that users are clear what they are expected to do.

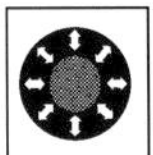

During the construction projects, many users did not understand what their roles and responsibilities were. Users need to be made aware of what they are expected to do or else they will not provide the correct information.

The users seem wary of making too many comments. They know that the person who has developed the system is sitting beside them and they do not want to say anything out of line. However, this defeats the purpose of the testing sessions. Also the developer does not encourage the users to say what they really think. The researcher finds it necessary to explain that constructive feedback is needed to improve the system and asks questions about specific aspects. The users now understand why they are needed and enter into proper debate.

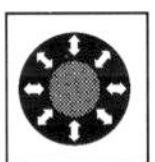

Similarly in the construction projects users were not always happy to speak out in front of their managers. It would perhaps have been better if the system could have been tested away from the developer.

In order to test the system, users are having to make up figures on the spot, but they are thinking on their feet and so their suggestions do not always tie up. Hence, they feel unable to comment on the accuracy of the system. It is agreed that it would have been sensible for them to have been better prepared and to have brought in details of a real project so that they could test the system in a knowledgeable manner.

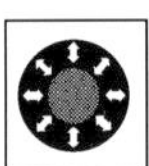

These users, like other users in the case studies, were unprepared for their relevant activities. In both situations the purpose of the session/meeting had not been properly explained and so users were having to make decisions on the spot which would often need to be revised later.

On several occasions the testing sessions are cancelled or cut short due to other ('real') commitments.

User reaction to the computer-based briefing system

When the different participants test the program for user friendliness, they also discuss the potential impact such a computer-based system could have on the briefing process. Their comments are summarised below:

Advantages

- The system provides the organisation with an innovative tool which will hopefully give them an advantage over their competitors if it is marketed properly.

- The amount of information required by the program before it arrives at an estimate for the cost of a scheme means that the client/design team are forced to consider certain decisions earlier on than they normally would. This means that sort sort of decision must be taken and cannot be ignored. Thus the program facilitates improved decision making.
- The level of detail required by the system means that it can be used as a briefing checklist to ensure that all of the necessary questions have been covered.
- Changes, or alternatives, can be input directly, rather than consultants having to go away and work things out. Hence, the system has the potential to speed construction projects up. Also people can instantly see the knock-on effects of certain decisions in relation to the overall project. Costs may decrease in one area, but increase somewhere else. In previous projects the impact of changes in one area often had an unforeseen cost, or other, implication elsewhere.
- The system allows each of the different scenarios to be saved. Hence, the organisation could build up a database of standardised information for specific building types. Thus the same basic information would not have to be generated each time.
- If the system is to produce sensible output it helps if the different members of the design team work together, each providing the answers in their relevant field. Thus the system fosters group decision making.

During the case studies and the implementation projects many of the problems and delays were caused by a lack of decision making. The system forces decisions to be made, otherwise it cannot produce output.

In traditional projects much time is wasted as consultants have to re-evaluate the project after changes are made to the design. With this system the implications of any decisions can be seen instantly and so consultants do not waste time on options that are obviously not feasible.

This would perhaps be particularly useful for clients who build similar projects on a frequent basis.

The system encourages team members to work alongside each other and so they work more as a group rather than individuals.

Disadvantages

- There is mistrust concerning the accuracy of the system. In order for the system to be able to make instant calculations it obviously contains standardised data. The figures or rates used do not always appear to correspond with the figures that consultants would use. The program needs to be able to accommodate the different rates used by different people.
- External consultants may view the system as a threat because potentially the program could make certain decisions on their behalf since it contains so much data

For any such system to work it has to be trusted by the people who use it. The only way for the system to be

trusted is for people to see that it works. Hence, the system should be used alongside real projects initially to see how it compares to traditional methods of working.

The system will only ever be as good as the people who use it and relies to a certain extent on people's experience and judgement of the specific situation. Hence, it should never be used to totally replace human input.

already, and so their input may not be necessary at the feasibility stage.

- Consultants could potentially find themselves under pressure by clients to meet the figures suggested by the program.

Summary

This project illustrates just how difficult it is to try to introduce new ideas/technologies into organisations. Even though senior managers were supportive of the initiative they did not free up people's time to allow them to comprehensively test the system. As with the construction projects, take-up of new ideas is extremely hesitant, the traditional methods still seem to be working, so organisations do not want to make changes – why rock the boat? It appears that it is only when difficulties arise that alternatives are even considered. The only way that this system (or any other innovation) is likely to be adopted is to use it in conjunction with well-tried methods and introduce it on an incremental basis.

12 Conclusions

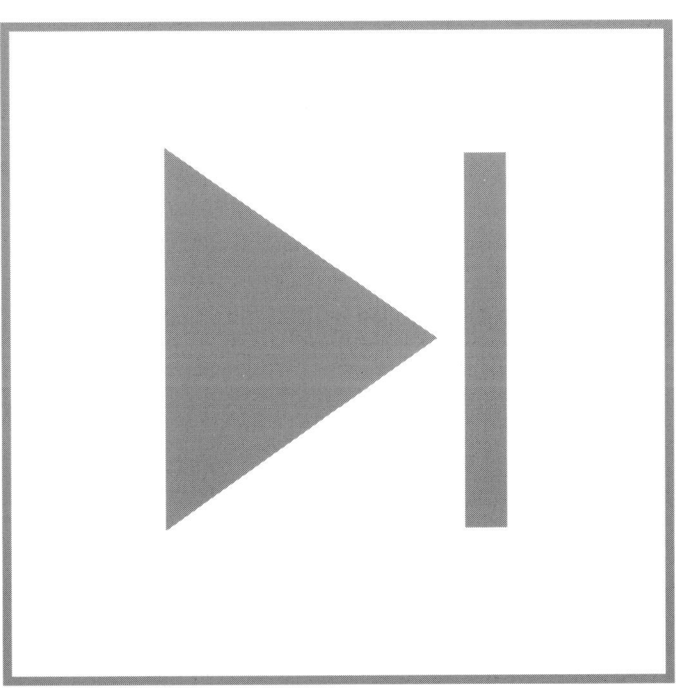

12 Conclusions

Good briefing, good business?

The advice given so far in this book has been orientated very much on the briefing process. However, just as to improve briefing an aspect has to be chosen and focused upon, at least initially, it is fair to ask would following the advice here not only improve briefing performance, but also overall business success through client satisfaction?

Would following the advice here not only improve briefing performance, but also overall business success through client satisfaction?

In general the thrust has been that the briefing process has to be continuous and highly interactive, with clients and construction professionals working together and complementing each other in terms of knowledge-base and interests. At this point in the book, before summarising the main advice for better briefing, we think it is interesting to compare this general approach with findings on effective marketing for construction professionals.

Work looking at exemplary construction consultancies has shown some interesting resonance with the above approach. Tony Faulkner (1996) carried out a detailed study of six consultancies selected by asking three very experienced clients to identify the two most exemplary practices they each dealt with. For each consultancy two projects were then studied in detail and a concerted effort was made to understand what made the firms exemplary. A common theme that became clear was the firms' ability to contribute to the decision-making processes as active partners with the clients. Alternatives were fully debated and joint problem-solving was the signature of the relationships. When problems arose, or mistakes were made, these were taken to the client openly and in a very positive spirit, with suggested solutions and all parties working together to resolve the issues. Overall there was an emphasis on service, not servility. The relationships were real as distinct from superficial. The

Faulkner, A.J. (1996) *Achieving Exemplary Quality in the UK Construction Professions*, unpublished MPhil thesis, University of Salford, UK.

firms' ability to contribute to the decision-making processes as active partners with the clients

service, not servility

firms were not simply responsive but made a 'generative' contribution (Senge, 1990), that is, they made a creative input that opened up new possibilities.

Senge, P. (1990) *The Fifth Dimension: The Art and Practice of The Learning Organization*, Doubleday, USA.

If this highly interactive approach is seen, at least by experienced clients, as exemplary, then what is the knock-on effect for the firm's business? From various studies it would appear that around 65% of construction consultants' new business comes from the existing clients in terms of repeat business. Thus, satisfying existing clients is very important for future workload. Of the remaining 35% around 61% comes from recommendation or referral as shown in Figure 12.1. Those recommending or referring are predominantly (around 81%) other clients or other professionals with whom the firm already works (see Figure 12.2). And, as summarised in Figure 12.3, the referral chains are mainly only one or two links long and the main media used are informal, that is, telephone or face-to-face conversations. Many people in construction use the rule that 'you are only as good as your last job'. The research by Hoxley (1993) appears to strongly back this up. If, even approximately, 65% of the consultancy's work is repeat business and of new work, 61% comes by recommendation and referral (81% of this through existing clients or other professionals with whom the firm works) then 82% of the firm's new commissions come via people who know the firm from first-hand experience.

Hoxley, M. (1993) *Obtaining and Retaining Clients: A Study of Service Quality and the UK Building Surveying Practices*, unpublished MPhil thesis, University of Salford, UK.

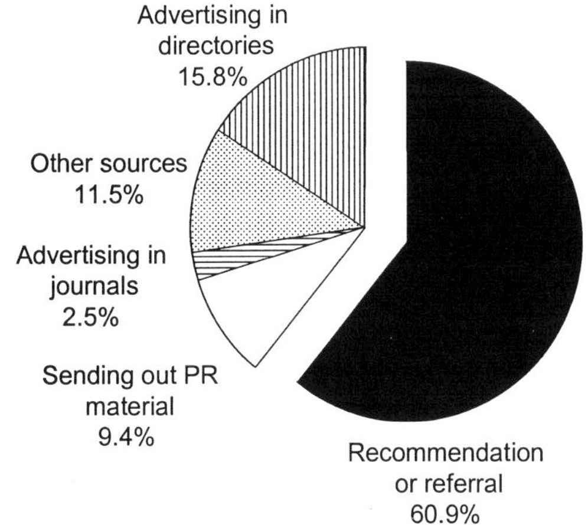

Figure 12.1
Firms' sources of new (non-repeat) work. Source: Hoxley, 1993

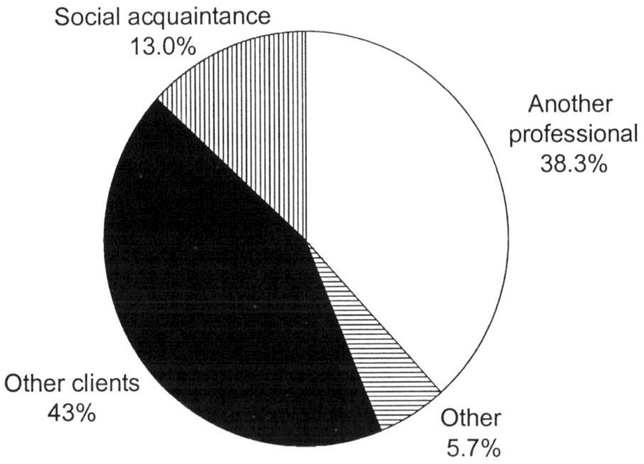

Figure 12.2
Firms' recommenders or referrers.
Source: Hoxley, 1993

Figure 12.3
Nature of referral chains.
Adapted from Hoxley, 1993

Length of referral chain		**Media used to refer**	
One link	65%	Telephone	66%
Two links	17%	Face to face	23%
Three links	6%	In writing	11%
Four links	2%		

The referral chain is short and informal

this demands 'real relationships'

The implication is that to succeed a firm must do a good job now. From the work on exemplary practices, this demands 'real relationships'. Briefing is the key interaction with clients through which their needs and objectives are interpreted into the construction project. If this can be achieved better by making the relationship work really positively, then there is every reason to believe that it will translate into greater business success.

good briefing is good business

In short, good briefing is good business.

The section below summarises the main features necessary to achieve progressively better briefing.

Summary

Chapter 1 started by exploring the question of why briefing practice did not appear to have improved significantly over the last thirty years, despite quite considerable best practice advice. It was found that such advice tended to be overly rationalistic to address many of the people issues involved and insufficiently flexible to allow firms to really take ownership of the ideas. The thrust of this book on *better briefing* was thus justified. We can't tell hands-on clients and professionals how to do briefing at a detailed level, but we can highlight common problem areas and present alternative approaches to overcoming them.

The second part of Chapter 1 therefore focused on key solutions areas, represented as a five-box model. It is suggested that if construction participants want to improve their briefing performance they should concentrate on: empowering the client, managing the project dynamics, appropriate user involvement, appropriate team building and appropriate visualisation techniques.

key solution areas, represented as a five-box model

Chapters 2–6 elaborated each of the five areas listed above. In each case the main ideas have been set out with some theoretical models and then clear, simple, practical advice given. This advice runs in parallel with live case study examples in the margin. We hope that the combination of ideas and examples will give construction participants a range of perspectives so that they will be motivated to try out some of the approaches themselves where they seem to make sense in their situation.

We do not underestimate the difficulty firms have in implementing changes when they are part of a set of complex relationships with other industry players. In Chapter 7 we have therefore explicitly discussed the question of implementation and provided a model of an incremental process. We do not necessarily expect firms to follow the process – we do expect them to experience it! Thus, the idea is that, as setbacks are experienced, firms will know that this is usual and keep their motivation to follow through. In a sense the message is one of lowering initial expectations, but it is linked to a strong enjoinder to maintain a sustained effort in order to achieve ambitious goals in the long term. Chapter

do not underestimate the difficulty firms have in implementing change

7 also illustrates the subtle point that the key improvement areas are systemically linked, thus it does not really matter where one starts, provided that a start is made.

Chapters 8–11 bring together many of the strands in four case studies. The central material is made up of the stories of the efforts of four construction firms to achieve improvements in briefing practice. In parallel text the linkage to the five key solution areas is drawn out in a narrative. These chapters illustrate, with practical examples, both the implementation process and the interactive nature of the issues involved.

In summary, the thrust of this book is that:

- Briefing performance needs to improve, but current best practice advice is of limited utility to those caught up in the hectic reality of briefing.
- Construction participants are encouraged to focus on five key improvement areas in order to improve their particular approach to briefing.
- Achieving change takes time and a sustained effort. It should start with something small but important to the firm.
- Many examples of alternative practical actions are given so that firms can choose what makes sense to them and easily use it.

Summary self-assessment

Against the above context, this is a suitable point to draw together the various self-assessments provided at the ends of Chapters 2–6. These exercises help identify priority factors for improvement action in each of the 'five-box' areas, based on an assessment of the major factors that had been highlighted in each area through the detailed assessment of many construction projects.

Table 12.1 provides a tool to create an overall summary of the priority factors for actions to improve briefing practice. Remember the tool can be used for the reader's organisation in general, some part of it, or it could be used to assess a specific project and just those people involved. For the summary to make sense, the assessments from the chapters should represent a consistent viewpoint.

Table 12.1
Overall summary of priority factors in all 'five-box' areas

Instructions
Drawing from the five summary analyses at the ends of Chapters 2–6, this table provides a tool to create an overall summary of key priority factors.

The factors located in white boxes in the chapter summaries are the highest priority for action, light grey the next and dark grey the lowest, given that entries in the black boxes are to be ignored as before.

Each of the factors listed below can, thus, be allocated with a 'cross' to one of the three levels of priority. By collecting the priorities in this way it should be possible to devise high leverage actions that will positively address more than one factor.

Factors	Highest priority (white area)	Moderate priority (light grey area)	Lower priority (dark grey area)
The summary analysis for 'empowering the client' is on page 39			
E1. Clients should be knowledgeable about their own organisations			
E2. Clients should be aware of the project constraints			
E3. Clients should understand the basics of the construction process			
E4. Clients should understand their roles and responsibilities			
E5. Clients should maintain participation in projects			
E6. Clients should gain the support of senior managers for projects			
E7. Clients should appoint internal project managers to manage projects			
E8. Clients should integrate business strategy and building requirements			
The summary analysis for 'managing the project dynamics' is on page 57			
D1. Teams should establish project constraints at an early stage			
D2. Teams should establish programme, highlighting critical dates			
D3. Teams should agree procedures and methods of working			
D4. Teams should allow adequate time to assess client's needs			
D5. Teams should validate information with the client organisation			
D6. Teams should improve feedback to all parties throughout project			
The summary analysis for 'appropriate user involvement' is on page 71			
U1. User involvement benefits should be understood			
U2. User involvement should be assessed relative to each situation			
U3. User participation should be planned to allow relevant data to be collected			
U4. User group dynamics should be considered			
U5. User involvement should be maintained throughout project			
The summary analysis for 'appropriate team building' is on page 81			
T1. Team selection should focus on skills, not just financial considerations			
T2. Teams should approve and understand management structure of projects			
T3. Team members should remain consistent throughout projects			
The summary analysis for 'appropriate visualisation techniques' is on page 90			
V1. Visualisation techniques should be employed to increase potential for shared understanding			
V2. Visualisation techniques should be adequately resourced			
V3. Visualisation techniques should be used effectively			

When Table 12.1 is complete it is easy to see the grouping of factors into the different priority bands. This provides the basis for designing action that synergistically links together high priority factors with issues that are important and urgent for the firm.

Final conclusions

Briefing was defined at the start of this book as:

The process running throughout the construction project by which means the client's requirements are progressively captured and translated into effect.

This is a broad conception, but it must be right that the client's requirements infuse the whole project. Equally, a recurrent theme has been that the relationship should be highly interactive so that the outcome is more than the sum of the parts. Opportunities are created and problems are dissolved through ingenuity. Something better than any of the individual parties to the project could have envisaged is achieved. Aesthetics are not sacrificed to function; costs do not suffer to achieve time demands; efforts to satisfy environmental criteria are not swamped in health and safety imperatives. To achieve this, improvement in briefing performance is vital.

The focus provided by the five key improvement areas, together with the implementation model and copious example actions, is a serious attempt to help individual firms take action. We expect these actions to be on a small scale to begin with, but over time there is evidence that the impact of the improvements will grow. Further, the improvement areas have been chosen so that the performance of the whole project coalition is enhanced, not only the instigator. Applied generally we are convinced that the performance of the whole construction industry will be greatly enhanced with progressively greater ease as more players orientate in the same direction.

the performance of the whole construction industry will be greatly enhanced with progressively greater ease as more players orientate in the same direction.

We are very optimistic that this general trend will come about, as the ideas have been presented in a way that is designed to facilitate practical take-up. Furthermore, briefing as defined is so central to the client relationship that good briefing is clearly good business. Firms may be unsure how well they carry out briefing at the moment, but they can be sure that better briefing will lead to better business.

We hope that firms will be encouraged to start improving their briefing **now**!

Appendix A – Further reading

The following references contain practical information and checklists which can easily be utilised by construction professionals and clients alike.

F. Salisbury, *Briefing Your Architect*. Architectural Press, Oxford, 1998.

> Clients, particularly inexperienced ones, will find this comprehensive book an invaluable guide to the briefing process. It explains which construction professionals should be involved in briefing and who should participate within the client organisation. The different stages of the briefing process are covered in detail and the tasks for each party at each stage are defined. Checklists, including the contents of an ideal brief, are provided throughout the book to ensure that all relevant work is carried out as the project develops. Clients who refer to this book will have a better understanding of what briefing entails, so will be empowered to participate fully in the process and are more likely to end up with a building that suits their needs.

Construction Industry Board, *Briefing the Team – A Guide to Better Briefing for Clients*. Thomas Telford, London, 1997.

> The aim of this short guide (20 pages) is to assist clients to improve the way they brief so that the design team fully understands their requirements. The following key stages of briefing are identified: statement of need, options appraisal, strategic brief, project evaluation plan, project brief. Each stage is supported by checklists of questions which clients should ask at each stage. This guide is part of a co-ordinated set of documents aimed at the improvement of processes within the construction industry. Clients may also find it useful to refer to

other documents in the series; for example, *Constructing Success: Code of Practice for Clients of the Construction Industry*.

D. Kernohan, J. Gray, J. Daish with D. Joiner, *User Participation in Building Design and Management*. Butterworth Architecture, Oxford, 1992.

This book focuses on the users of buildings. The authors discuss why and how users should be consulted during the building process. They advocate the use of a simple generic building evaluation process that encourages dialogue between users and building designers. Various case studies illustrate how their generic evaluation process has been successfully used to highlight user concerns and preference in real life projects. Step-by-step advice is provided for construction professionals on how to undertake such building evaluations. Users who are planning a move or a refurbishment will find it helps them to think about their needs in a logical manner and evaluate their existing surroundings so that they can pass on relevant information to the designers.

Appendix B – Functional brief

Information in pro forma adapted from: Becker, F. (1990) *The Total Workplace*, Van Nostrand Reinhold, New York.

Clients and construction professionals may find this pro forma useful as a prompt when considering briefing requirements. However, the list is by no means exhaustive. The right hand margin is blank so that you can photocopy the list and add your comments. Organisations who are involved with construction projects on a regular basis may wish to adapt the list to suit their particular circumstances.

Project overview	
Project stakeholders Who are the major stakeholders within the client organisation? Which other people are involved in the project? How will the project management be structured? Who should be consulted during briefing? Who is authorised to make which decisions?	
Aims and background What does the client hope to achieve? Why has this project come about?	
Fixed constraints What decisions are fixed before briefing starts? What finances are available for the project? What is the intended time span of the project?	
Organisational concerns	
Organisational structure/culture What is the structure of the organisation? How will people interact with one another? How will people use their time/space? Will people work in the building outside of normal office hours?	

Staff How many people will the building accommodate? Is this likely to alter over time? What sort of staff will occupy the building? What mix is there likely to be of the sexes?	
Image expectations What image should the building present to the outside world?	
Individuals and workstyles	
Task analysis What exactly will different members of staff do?	
Environmental satisfaction What sort of work environment will staff require to carry out their jobs?	
Communication and adjacency Who will communicate with whom? How, where, when and how often? Which people/groups need to be located in close proximity?	
Space, furniture and equipment requirements What equipment will be used in the building? What furniture and equipment is necessary in order for people to perform their jobs? How much space is needed for different tasks? Will existing equipment or furniture be re-used?	
Physical environment	
Security Which areas will require public access? Which areas need to be secure? What about after-hours access? Are different levels of security required?	

Circulation requirements How will people move through the building? What are the requirements for lifts, staircases and ramps? What access is required for disabled people? How will equipment be moved around?	
Transportation and parking How many staff car parking spaces are needed? How many spaces are required by visitors? What is public transport like in the area? How often do deliveries take place?	
Surrounding amenities What facilities are provided in the immediate area that do not need to be replicated? For example, are there local facilities for food?	
Appearance Does the client have any preferences relating to the look of the building? For example, form, scale, choice of materials, colour, proportion – both external and internal.	
Inspirations Does the client (or design team) know of any other buildings which are used for similar purposes? Are there other buildings which are used for alternative purposes that may provide inspiration?	
External influences	
Laws and codes Is there any specific legislation that needs to be considered, other than construction related?	
New technologies Are any changes likely to happen over the next few years that will have an impact on the way the building is designed?	

Other users Which other people will visit the building? What resources will they require? Which areas will they have access to?	
Competitors and interested parties How do the proposals compare to competitors' buildings? Are there any interested parties who may be able to provide useful input to the design?	
Specialist considerations	
Environmental policies Does the client have any specific environmental requirements or policies?	
Heating and cooling requirements Are there any pieces of equipment/areas that have specific heating/cooling requirements? Will users want localised temperature control? WIll this conflict with other requirements?	
Acoustics Are there likely to be any areas that have specific acoustic requirements?	
Lighting Which areas require natural lighting? Will any areas require unusual lighting levels? What lighting levels will staff activities require?	
Contamination protection Will equipment/areas have specialist requirements? For example, ventilation, humidity, protection against dust and dirt?	
Loading requirements Will there be any unusually heavy equipment or loads?	
Energy levels What energy levels will be required? Will an alternative supply be necessary?	

Maintenance needs and running costs What is the estimated lifespan of the building? Who will be responsible for maintenance? How important will running costs be? How will refuse be stored and removed? What areas will have specific cleaning requirements?	
Add your own questions here	

Appendix C – Description of the research basis of the book

Background

The research project which provided the foundation for the recommendations given in this book was carried out over two and a half years from 1995 to 1997. Funding was provided by the Engineering and Physical Sciences Research Council (EPSRC) and the Department of the Environment, Transport and the Regions (DETR) through a research programme called LINK IDAC. This provided the support necessary for research assistants, whilst industry provided matching resources through their involvement. The industry partners were Aqumen Plc, Currie & Brown, Ernst & Young, Laing North West, Nuffield Hospitals and Property Tectonics.

Given the nature of the research programme, industry was very closely involved in every stage of the process, particularly the provision of data and commentary on the analysis carried out by the academic partners in the Research Centre for the Built and Human Environment at Salford University. The academic team proposed early on that an explicitly rigorous research methodology should be used in tandem with the extensive experience available and this was enthusiastically accepted by the partners and pursued consistently throughout.

Programme

The overall programme for the project is given in Figure C1. It can be seen that there are three main phases, namely a pilot case study phase, a main case study phase and an action research phase. In broad terms the pilot case study phase supported the work in Chapter 1 in which the nature and impact of 'best practice' was clarified. The main case studies

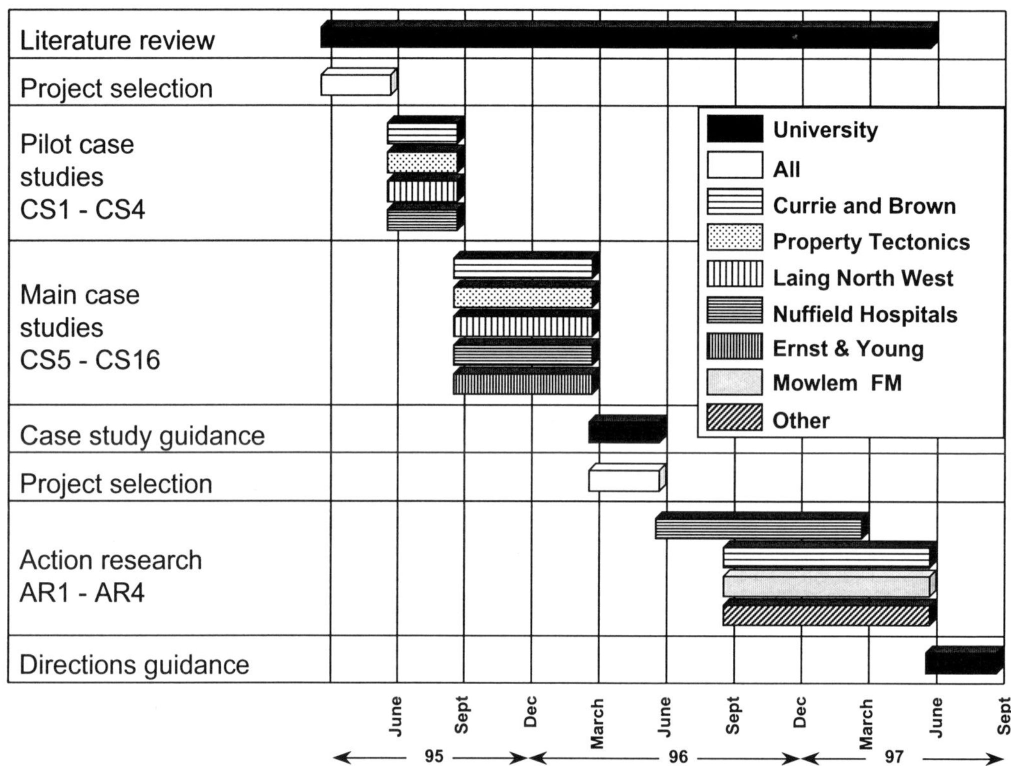

Figure C1
Project programme

gave rise to the five-box model introduced at the end of Chapter 1 and elaborated, drawing extensively on the case study material, in Chapters 2–6. The action research phase supported the development of the implementation model described in Chapter 7 and illustrated in Chapters 8–11.

Grounded theory methodology

The project was predicated on unsatisfactory progress over thirty years, despite a lot of good practice advice. Thus, it was decided early on that recycling of the same ideas was unlikely to lead to new insights. This pointed very clearly towards a project methodology that dealt with rich data drawn from current practice. In this way there would be opportunity to understand why good practice results had not impacted as strongly as would be expected so far and a wealth of illustrative material would be generated in the process. The main underpinning methodology chosen was, therefore, drawn from grounded theory (Glaser and Strauss, 1967). The emphasis is therefore on the generation of new theory, rather than the validation of old ideas. To support this, variety in the data is needed, rather than a random sample that will tend to concentrate on the middle ground and overlook interesting exceptions. Figure C2 shows the practical outcome of this approach, where variety was sought in the projects studied. The projects ranged in value from £200 000 to £10 million; some are new build, some refurbishment; they cover different procurement methods, a variety of building types and are at a number of different stages of development varying from inception to near

Glaser, B.G. and Strauss, A.L. (1967) *The Discovery of Grounded Theory*, Aldine, Chicago.

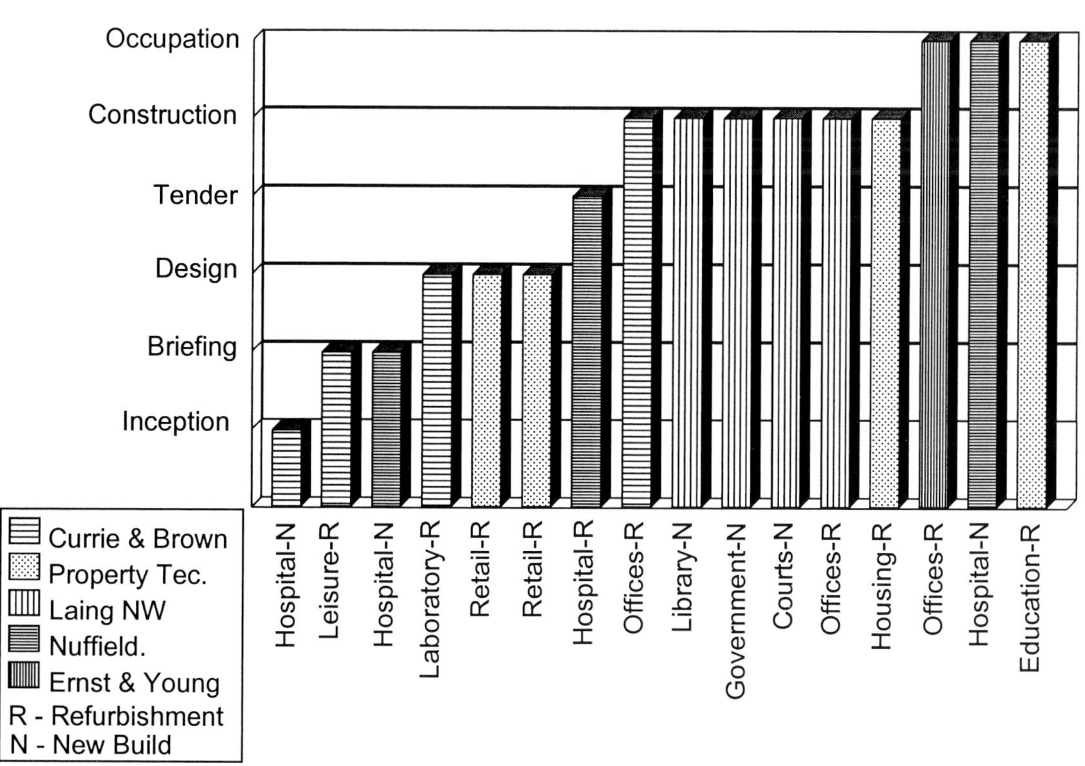

Figure C2
Overview of case studies

completion. Several participants have been interviewed for each project in order to gain an increased understanding of the different issues involved. This multiple perspective aspect of the project draws from soft systems ideas (e.g. Wilson, 1984) and further triangulates the work.

Wilson, B. (1984) *Sytems: Concepts, Methodologies and Applications*, John Wiley and Sons, Chichester.

Strauss, A.L. and Corbin, J. (1990) *Basics of Qualitative Research: Grounded Theory Procedures and Techniques*, Sage, Newbury Park, California.

Based on the review of the methodological literature (Strauss and Corbin, 1990) the broad approach illustrated in Figure C3 was defined. This has various important features which will be discussed below, working from the bottom of the pyramid up. In order to mitigate some of the weaknesses of studies based on soft data analysis, the first action was to define some consistent case study procedures for the analysis. A decision was taken that the main data collection source would be interviews and that these would be recorded and translated into near verbatim records. These were then sent to interviewees for checking to ensure that misinterpretations had been avoided (internal validity). These transcripts were designed to carry as much information as possible and in practice this meant that a margin space was provided for commentary by the researcher involved and within the text different notations were used to carry extra information. For

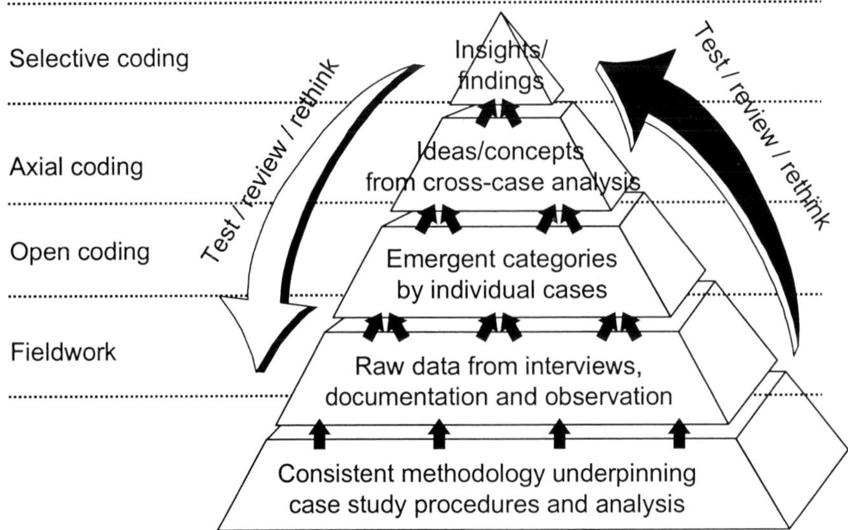

Figure C3
Grounded theory approach

example, bold type was used to indicate emotion in the interviewee, italics were used to indicate opinions and references to information sources were underlined.

The various transcripts taken together make up the second slice of the pyramid, termed 'fieldwork' in Figure C3. A key aspect of the approach is to keep this separate, as raw data, so that it can both support analysis and still be available for re-analysis later in the project. This latter facility is important as 'theoretical sensitivity' increases as the project proceeds and additional insights become possible from the data as the experience of those involved increases.

Identifying a category of record as raw data prompted the identification of other categories of record. First, there are the analyses of the raw data that permeate the upper three levels of coding and, second, there are records recording decisions and thoughts about the research design and execution itself, which again are kept separately.

It can be seen that a lot of data was involved and early on a decision was taken to use software to support the organisation and analysis of various records. The well known NUD.ist program (**n**on-numerical, **u**nstructured **d**ata, **i**ndexing, **s**earching, **t**heorising) was selected and an information hierarchy created as shown in Figure C4. Thus, underpinned by standard procedures, and supported by flexible software, raw data was collected and organised and then analysed (open coding) until ultimately the progression from emergent categories to more general ideas and concepts could be turned into insights and findings at the top of the pyramid. At this stage the data can then be used 'selectively' to illustrate the findings and resolve any remaining inconsistencies.

NUD.ist software is distributed by Sage Publications Ltd, 6 Bonhill Street, London EC2A 4PU.

Practice process followed

The research methodology described above did stand up well in practice. However, the impression should not be given that progress was straightforward and linear. In practice the methodology supported flexibility and progress was incremental and iterative. The idealised image of the process

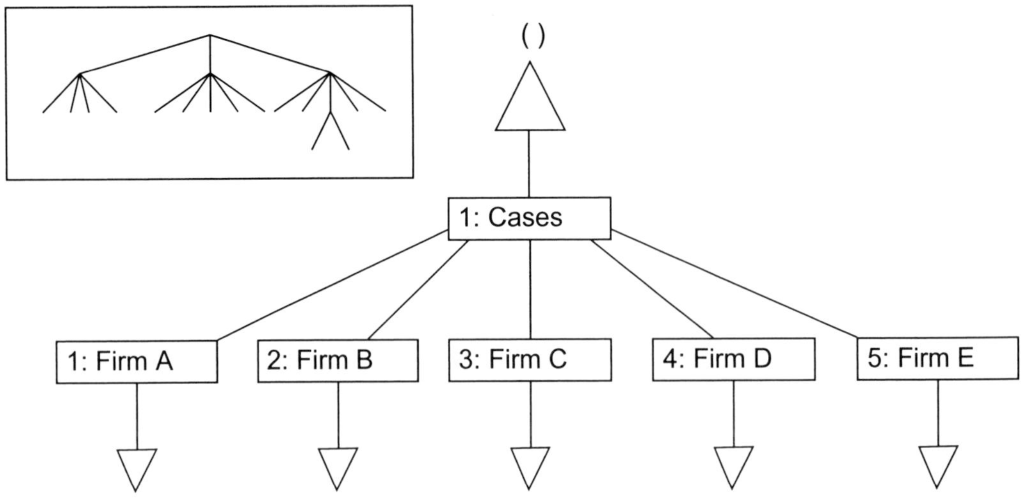

Figure C4
Example information tree

would be of an upwards spiral. Various cycles were completed, during which a better level of understanding was achieved and then discarded once it had provided a foundation for the next level of analysis.

The project initially sought to identify typical problems associated with briefing and examples of best practice in the industry. It utilised a multiple case study approach (Yin, 1989). An initial pilot survey of five projects was carried out to test the adequacy of the survey methods and to identify the overriding issues which were of concern to different participants in the briefing process. Material for the main case studies was then gathered through interviews and analysis of documentation at different levels in the organisations involved. The data was being assessed at the level of holistic cases (projects), embedded cases (incidents within projects) and through cross-case comparisons at both of these levels. In total sixteen projects were considered during the pilot/main case studies phase.

Yin, R. (1989) *Case Study Research – Design and Methods*. Sage Publications, Newbury Park, California.

The resulting 500 pages of data were analysed within cases by two researchers working in parallel and then resolving the resulting problem categories each had produced (closed coding). This led to a combined long list of categories identified by keywords. These were then analysed for prevalence using the software and the eight most common problem areas identified. The data were then searched and problems and their contexts assessed around the eight primary problem areas (axial coding). At this stage a creative leap was needed to reach the insights desired. So, cognitive maps were used (see Figures 1.4–1.11 as examples) and possible solutions were added by induction on the part of the research team, although clearly the researchers drew from their familiarity with the material. This device then allowed a synthesis of the eight maps down to five major solution areas that were then used to re-collect the raw material together into a number of major factors for each of the five areas (selective coding).

At this stage of the analysis, the project deviated in an important respect from a 'pure' grounded theory approach. Whilst maintaining the benefits of a raw data reservoir, steps were taken to avoid the limitations of relying on insights *only* from this data and the researchers' views of it. Strauss and Corbin (1990) advise a 'process' paradigm for the axial coding stage. This represents *one* 'world view' or Weltanshuung (W), using soft systems terminology (e.g. Wilson, 1984). We extended the multiple perspectives approach further by combining emergent theory from the grounded approach with the stimulus from a wide range of Ws created by a diverse range of *existing theories*. All were tested back against the raw data to ensure coherence and the number and diversity of focused analyses mitigates against the danger of any one existing theory blinding the researchers. Theoretical stimulus has been drawn from areas such as negotiation, leadership, organisational growth, socialisation, human error, group dynamics and decision-making.

The action research phase allowed the ideas to be tested in real world situations. The same research tools enabled these experiences to be recorded and analysed with a degree of rigour. This led to some improvement in the findings and many illustrations. It also opened up a big area of interest as to how the ideas were used as well as what the ideas were. This implementation issue arose at the same time as an international workshop on this topic and this is a specific area where others' theories helped make some sense of what we were observing first hand in our research.

NBI Workshop CIB W65 (Organisation and Management of Construction), *Transfer of Construction Management Best Practice Between Different Cultures*, 6–7 June 1997, Oslo, Norway.

Summary

The project methodology was chosen to allow new insights to emerge that would be relevant at an industry level. This argued for a grounded theory approach in which data were elicited through case studies designed to provide a rich variety of perspectives. The research process was underpinned by well defined protocols and flexible software support. This allowed an incremental, iterative approach to be pursued, through which new insights were ultimately created and illustrated and then linked and infused with relevant existing theory.

Evaluation

It is appropriate at this stage to briefly reflect on whether this qualitative research has achieved rigour, although it is, of course, for others to reach objective conclusions on this.

Guba, E. and Lincoln, Y. (1989) *Fourth Generation Evaluator*, Sage, Newbury Park, California (quoted in Symon, G. and Cassell, C. (1998) *Qualitative Methods and Analysis in Organisational Research: A Practical Guide*, Sage Publications, London, p. 7).

Guba and Lincoln (1989) suggest the following criteria for assessing the 'authenticity' of qualitative research: resonance, rhetoric, empowerment and applicability.

Resonance is the extent to which the research process reflects the underlying paradigm. The topic is highly subjective and involves many participants interacting through complex relationships. The combination of case studies followed by synthesis through cognitive mapping is resonant with these characteristics.

Rhetoric is concerned with the strength of the argument presented. By linking multiple views to synthesise the major issues around five key areas a strong case has been provided for making improvements in briefing practice.

This links to *empowerment*, that is, enabling readers to take action. There is a clear agenda, but it remains to be seen if these findings, illustrated as they are, will lead to concerted action in a combative industry.

The explicit consideration of the process of implementation should help, together with the high *applicability* of the findings, achieved through keeping visible the local roots of the findings. It seems from industry feedback that everyone can find something in the material that confronts them in their everyday lives.

References

Asch, S., 'Opinions and Social Pressures', *Scientific America*, November, 1955, pp. 31–5.

Barrett, P.S. (ed), *Facilities Management: Towards Best Practice*, Blackwell Science, Oxford, 1995.

Bejder, E., 'From client's brief to end use: the pursuit of quality', in *Practice Management: New Perspectives for the Construction Professional*, P.S. Barrett and A.R. Males (eds), E&FN Spon, London, 1991, pp. 193–203.

Bono, E. de, *Lateral Thinking for Management*, McGraw-Hill Book Company (UK) Ltd., 1971.

Construction Industry Board, *Briefing the Team – A Guide to Better Briefing for Clients*, Thomas Telford, London, 1997.

Faulkner, A.J., *Achieving Exemplary Quality in the UK Construction Professions*, unpublished MPhil Thesis, University of Salford, UK, 1996.

Garrett, R., 'Facing up to change', *Architects Journal*, 28/10/1981, pp. 838–42.

Gelb, M.J. and Buzan T., *Lessons from the Art of Juggling*, Aurum Press (1995), London, 1994.

Glaser, B.G. and Strauss, A.L., *The Discovery of Grounded Theory*. Aldine, Chicago, 1967.

Guba, E. and Lincoln, Y., *Fourth Generation Evaluator*, Sage, Newbury Park, California, 1989 (quoted in G. Symon and C. Cassell, *Qualitative Methods and Analysis in Organisational Research: A Practical Guide*, Sage Publications, London, 1998, p. 7).

Handy, C.B., *Understanding Organisations*, Penguin, Harmondsworth, 1985.

Hersey, P., Blanchard, K.H. and Johnson, D.E., *Management of Organisational Behaviour: Utilising Human Resources*, 7th edition, Prentice-Hall, New Jersey, 1996.

Hoxley, M., *Obtaining and Retaining Clients: A Study of Service Quality and the UK Building Surveying Practices*, unpublished MPhil thesis, University of Salford, UK, 1993.

Kernohan, D., Gray, J., Daish, J. with Joiner, D., *User Participation in Building Design and Management*, Butterworth Architecture, Oxford, 1992.

Kolb, D.A., *The Learning Style Inventory Technical Manual*, McBer, Boston, 1976.

Latham, M., *Constructing the Team* (The Latham Report), HMSO, London, 1994.

Lewin, K., *Field Theory in Social Sciences*, Harper and Row, New York, 1951.

Luft, J. and Ingham, H., 'The Johari Window, A Graphic Model of Interpersonal Awareness', in *Proceedings of the Western Training Laboratory in Group Development*, UCLA Extension Office, Los Angeles, 1955.

Mackinder, M. and Marvin, H., *Design Decision Making in Architectural Practice*, Institute of Advanced Architectural Studies: Research Paper 19, University of York, York, 1982.

Maister, D.H., 'Balancing the Professional Service Firm', *Sloane Management Review*, MIT, Fall, 1982, pp. 15–29.

March, J.G. and Olsen, J.P., 'The Technology of Foolishness', in *Organization Theory*, D.S. Pugh (ed), 4th edition, Penguin, London, 1997, pp. 339–52.

Megginsson, L.C., Mosley, D.C. and Pietri, P.H., *Management Concepts and Applications*, 3rd edition, Harper and Row, Cambridge, Mass, 1989.

Ministry of Public Buildings and Works, *The Placing and Management of Contracts for Building and Civil Engineering Work* (The Banwell Report), HMSO, London, 1964.

Minzberg, H. and Waters, J.A., 'Of strategies deliberate and emergent,' *Strategic Management Journal*, Vol 6, 1985, pp. 257–72.

Newman, R., Jenks, M., Bacon, V. and Dawson, S., *Brief Formulation and the Design of Buildings*, Oxford Polytechnic, Oxford, 1981.

O'Reilly, J., *Better Briefing Means Better Building*, BRE, Garston, Watford.

Pena, W.M., Caudill, W. and Focke, J., *Problem Seeking: an architectural programming primer*, Cahners, Boston, 1977.

Powell, J.A., 'Clients, designers and contractors: the harmony of able design teams', in *Practice Management: New Perspectives for the Construction Professional*, P.S. Barrett and

A.R. Males (eds), E&FN Spon, London, 1991, pp. 137–48.

RIBA, *Plan of Work*, RIBA, London, first published 1967, regularly updated since.

Reason, J., *Human Error*, Cambridge University Press, Cambridge, 1990.

Salisbury, F., *Architect's Handbook for Client Briefing*, Butterworth Architecture, London, 1990.

Salisbury, F., *Briefing Your Client*, Architectural Press, Oxford, 1998.

Senge, P., *The Fifth Dimension: The Art and Practice of The Learning Organization*, Doubleday, USA, 1990.

Sjoholt, O. (ed), *Transfer of Construction Management Best Practice Between Different Cultures*, (CIB W-65, Workshop Proceedings, Oslo, Norway), Publication 205, Conseil International du Bâtiment, Rotterdam, Netherlands, 1997.

Spekkink, D., *Programma van Eisen (SBR 258)*, Stichting Bouwresearch, Rotterdam.

Spekkink, D. and Smits, F., *The Client's Brief: more than a questionnaire*, Stichting Bouwresearch, Rotterdam, 1993.

Strauss, A. and Corbin, J., *Basics of Qualitative Research: Grounded Theory Procedures and Techniques*, Sage, Newbury Park, California, 1990.

Tufte, E.R., *Envisioning Information*, Graphics Press, Connecticut, USA, 1990.

Whyte, W., *Participatory Action Research*, Sage Publications, Newbury Park, California, 1991.

Wilson, B., *Systems: Concepts, Methodologies and Applications*, John Wiley and Sons, Chichester, 1984.

Yin, R., *Case Study Research – Design and Methods*, Sage Publications, Newbury Park, California, 1989.

Zeisel, J., *Inquiry by Design*, Cambridge University Press, Cambridge, 1984.

Index